....

**Selective Immunosuppression: Basic Concepts and
Clinical Applications**

....

Chemical Immunology

Vol. 60

Series Editors   *Luciano Adorini,* Milan
                    Ken-ichi Arai, Tokyo
                    J. Donald Capra, Houston, Tex.
                    Kimishige Ishizaka, La Jolla, Calif.
                    Herman Waldmann, Oxford
                    Byron H. Waksman, New York, N.Y.

KARGER   Basel · Freiburg · Paris · London · New York ·
            New Delhi · Bangkok · Singapore · Tokyo · Sydney

Selective Immunosuppression: Basic Concepts and Clinical Applications

Volume Editor   *L. Adorini*, Milan

22 figures and 7 tables, 1995

KARGER   Basel · Freiburg · Paris · London · New York ·
New Delhi · Bangkok · Singapore · Tokyo · Sydney

....

Chemical Immunology

Formerly published as 'Progress in Allergy'
Founded 1939 by Paul Kallòs

RC
583
.P7
V.60

Bibliographic Indices. This publication is listed in bibliographic services, including Current Contents® and Index Medicus

Drug Dosage. The authors and the publisher have exerted every effort to ensure that drug selection and dosage set forth in this text are in accord with current recommendations and practice at the time of publication. However, in view of ongoing research, changes in government regulations, and the constant flow of information relating to drug therapy and drug reactions, the reader is urged to check the package insert for each drug for any change in indications and dosage and for added warnings and precautions. This is particularly important when the recommended agent is a new and/or infrequently employed drug.

© Copyright 1995 by S. Karger AG, P. O. Box, CH–4009 Basel (Switzerland)
Printed in Switzerland on acid-free paper by Thür AG Offsetdruck, Pratteln
ISBN 3–8055–6034–6

Contents

MHC Blocking Peptides and T-Cell Receptor Antagonists: Novel Paths to Selective Immunosuppression?

Modified T-Cell Receptor Ligands: Moving beyond a Strict Occupancy Model for T-Cell Activation by Antigen

Idiotypic Regulation Directed at T-Cell Receptor Determinants

Selective Immunosuppression: Where Are We Now, and Where Are We Going?

A Preface

Selective immunosuppression is at the frontier of immunological research, and enough is known about the immune system to permit new ideas on immunoregulation to be tested in patients. Several concepts validated in experimental systems are currently being applied in clinical situations: a few pass, some fail, and most are encouraging enough to be submitted for further testing. This is a very interesting time to survey what is happening along the frontier, as sufficient information is now available on basic mechanisms and their clinical applicability to suggest the most promising pathways for future development.

Selective immunosuppression can be induced by a variety of approaches, which could be grouped into two broad categories: approaches directed specifically at autoreactive T cells by targeting the MHC/antigenic peptide/TCR complex, and less-selective approaches targeting a substantial fraction of T cells, including the pathogenic ones.

The first category includes very effective modes of immunosuppression, at least in experimental models. The holy grail is induction of tolerance to the autoantigens as a treatment for human autoimmune diseases. This requires, in principle, knowledge of the autoantigens, still poorly defined in most autoimmune situations, although progress is expected in their identification and characterization. Once the inciting autoantigen has been identified, specific immunosuppression of T cells recognizing it could be induced by exploiting one or more of the mechanisms controlling the peripheral tolerance described in this volume. In this respect, basic research has made considerable advances, but induction of tolerance to the relevant antigen in clinical situations still remains a long-term goal. This category also includes approaches targeting MHC molecules or the TCR. Unfortunately, MHC blockade can only prevent, not treat, autoimmune diseases. In addition, peptides, due to their unfavorable pharmacokinetics, cannot be developed as MHC antagonists. Approaches targeting the TCR itself are also very problematic; the TCR used by human pathogenic T cells is probably too heterogeneous to represent a useful target for immunosuppression. The problems surfacing in the clinical applicability of these approaches definitely represent major challenges for pharmacological development.

In the second category of approaches, very heterogeneous indeed, some strategies have been clinically tested more thoroughly. Among them, some look extremely promising, e.g. TNFα antagonists in rheumatoid arthritis. It is likely that the next generation of immunosuppressive drugs will include several cytokine antagonists, and in particular those able to inhibit, directly or indirectly, the development and function of Th1 or Th2 CD4+ T cells. This underscores the impact on immunotherapy of the current paradigm in Immunology: the Th1/Th2 dichotomy. The subdivision of T cells into Th1 and Th2 subsets can be oversimplified to suggest that most organ-specific autoimmune diseases are Th1 mediated, whereas immediate-type hypersensitivities are Th2 mediated. Although clinical situations are certainly more complex, this paradigm offers the possibility to design straightforward experiments to probe the role of Th1 and Th2 cells in immunoregulation and in the pathogenesis of immunological diseases. It is hoped that tipping the Th1/Th2 balance may offer novel approaches for immunointervention in autoimmune diseases and allergies.

In this volume, we are accompanied on the road from basic concepts to clinical applications of selective immunosuppression by leading immunologists, each with a distinct interest in applying the progress of immunological research to the treatment of human diseases. I think their contributions have assembled a very interesting volume, covering many facets and portraying the state of the art in selective immunosuppression. I would like to thank them for sharing their views and their thoughts with us. I also wish to acknowledge the editorial skills of Marianne Fratangelo, and her help in bringing this volume from an idea to a reality.

Luciano Adorini

Adorini L (ed): Selective Immunosuppression: Basic Concepts and Clinical Applications.
Chem Immunol. Basel, Karger, 1995, vol 60, pp 1–19

....

Self-Determinant Selection and Selective Regulation

Vipin Kumar, Eli Sercarz

Department of Microbiology and Molecular Genetics, University of California,
Los Angeles, Calif., USA

Introductory Comments

The presence of self-reactive T cells and self recognition is now considered essential for the normal functioning of the immune system. The ease with which self-reactive T cells can be detected in normal individuals indicates that negative selection is incomplete and that other mechanisms must operate subsequently to maintain self tolerance. In this review, we will consider the following mechanisms: (1) sequestration of self proteins as well as self determinants as a result of selective processing or tolerance; (2) containment (reduced number or activity) of self-reactive T cells by network regulation; (3) induction of preferentially nonpathogenic Th2-like cells. We will also describe the importance of understanding the mechanisms of nonresponsiveness to both self and nonself antigens in exploration of tolerance mechanisms.

Clonal anergy is considered one of the mechanisms important for peripheral tolerance [1]. Mature T cells become anergic if they are not appropriately triggered by professional antigen-presenting cells. T-cell anergy has been defined as the inability of T cells to produce interleukin-2 (IL-2) and to proliferate. To study tolerance to tissue-specific antigens that are not expressed in the thymus, tissue-specific promoters have been used to direct expression of mouse MHC class I, class II, or viral antigens in transgenic mice [2]. In some cases, T cells were rendered anergic; in other cases self tolerance was not induced. In the periphery, clonal deletion or 'exhaustion' can also occur after clonal expansion [3, 4]. Stockinger and Lin [5] found that macrophages from C5-deficient mice efficiently present their endogenous pro-C5 to activate CD4+ T cell clones in vitro. However, these pro-C5-reactive T cells remain functionally silent and are not anergized or activated in vivo. Self cryptic determinants lacking access to negative selection machinery owing to a lower level of expression or to a defect in the processing machinery, are essentially ignored. On the other hand, if small resting B cells which fail to activate T cells indeed provide negative signals under physiological conditions in vivo [6], they might be the prime candidate for inactivating T cells that have escaped thymic deletion. It may also be that T-cell activation follows different rules in the thymus than in the periphery.

Although the above-mentioned mechanisms have been proposed to maintain peripheral tolerance, their direct involvement in an autoimmune disease model has not been demonstrated. Therefore, we will discuss other possibilities including immune regulatory mechanisms that may be responsible for maintaining tolerance to antigens in the periphery. Maintenance of tolerance to self via lack of appropriate T-cell help has also recently been demonstrated for class I-restricted CTL reactive to extrathymic antigens, suggesting that self-nonself descrimination may be primarily carried out by CD4+ T cells. Moreover, CD4+ T cells with the capacity to down-regulate EAE [7–9] or collagen-induced arthritis [10] have also been demonstrated.

Self Determinants, Self Repertoire and Tolerance

Only a Few Self Antigens and Their Determinants May Be Involved in Autoimmune Pathology

It is likely that the immune system regulates predominant T-cell-receptor idiotypes that are associated with pervasive determinants of the 'immune homunculus' as proposed by Cohen and Young [11]. These highly represented determinants induce strong responses against which a regulatory apparatus has already been installed. It is likely that only a few V idiotypes in particular gene families are involved. In this situation, if a homuncular T cell expresses a particular Vβ gene family, then the homuncular determinant would be one that was readily processed and of high affinity for the MHC, i.e. dominant, and should invariably be able to induce self-regulatory cells. In order to induce responses to other determinants, the homuncular determinants could be chosen from a few Vβ families, and examples of such restriction in the use of certain gene families have already been described [12]. Regulation could also involve homuncular 'idiotopes' shared by different V gene families and here specificities might include both α- and β-chains. The rationale for the focus by the immune system on a very few determinants is still a matter of conjecture [11, 13, 14]. It is likely that a narrow antigenic focus leads to a response which is more cogently regulated.

We know little about the options available to the immune network in determining whether or how a particular response should be controlled. If an antigen induced both cellular and humoral responses, how and at what level would these two arms of the immune system interact? There are numerous examples of early regulation by mutually inhibiting lymphokines at the level of choice between Th1 and Th2 [15]. Certainly, the role of the immunoglobulin idiotypes in shaping the T-cell repertoire in the periphery [16] suggests that interactions between a processed idiopeptide from Ig and specific T cells induced against a determinant may be one of the determining factors in the ultimate outcome of an immune response. Regulation via a T-cell network seems to be an important alternative which can reflect a feedback mechanism based on awareness of the ambient cellular milieu.

Conserved Stress Proteins Might Be an Important Target in Autoimmune Pathology

Infection with bacteria or other cellular parasites exposes the immune system to a large number of foreign antigens simultaneously. It is surprising that hsp65 molecules from many different infectious agents are dominant antigens in spite of the fact that they are some of the most phylogenetically conserved molecules in nature. Thus, processing and presentation of self hsps occurs readily and T cells with specificity for self hsp exist. Such T cells can be activated during infection and autoimmunity. HSP determinants would represent a favored homuncular example, since such a widespread system of chaperones would constitute an excellent target for regulation.

NOD mice spontaneously develop IDDM and since T cells can transfer diabetes in these mice, it seems likely that the autoimmune process involves a T-cell response to a target pancreatic antigen. It has been shown that mammalian hsp65 is one such antigen. T-cell clones reactive to a determinant of human hsp65 not only can transfer diabetes but also vaccinate against it [17, 18]. This determinant is identical in sequence to the mouse hsp65 except for a single amino acid. What can be expected is that a primary diabetogenic determinant is the initial focus for a response which then spreads to other sites on the same molecule and then on to others [19]. It is also likely that since heat-shock proteins (hsp) perform an auxiliary and necessary function in many cells, disease could be perpetuated by hsp peptides after initiation by an organ-specific antigen. Therefore, hsp determinants could function as potentially useful targets to ameliorate some of the pathology in autoimmune diseases. It is also possible that a regulatory anti-idiotypic T-cell network could be organized around T cells specific for a few dominant antigenic motifs on hsp.

Self-Reactive T Cells Utilize a Limited Set of TCR Variable Genes

Studies on TCR variable gene usage for defined antigenic determinants are in general consistent with the hypothesis that pathogenic T cells are focused on very limited sets of peptide determinants and express a limited repertoire of TCRs [20, 21]. It has been demonstrated that there is extremely restricted Vβ gene usage during the T-cell response to SWM 111–121 (I-E^d restricted) or to MBP 1–11 (I-A^u restricted), with a single Vβ chain dominating both responses – Vβ8.2 [22, 23]. The majority (59%) of collagen-specific T cells also predominantly utilize the Vβ8.2 gene segment [24].

In studies on the heterogeneous TCR repertoire of cells infiltrating the islets in NOD mice [25], as well as in some reports in human systems, it is possible that the apparent TCR sequences are diverse because many of them are either not derived from autoreactive T cells, or certainly not from T cells that initiate autoimmunity. Some T cells may have migrated to the inflamed site under the influence of locally released cytokines, although they are not responsive to self antigens. Similarly, T cells reactive to other self antigens or determinants may be recruited later in the response. It may therefore be important to study the TCRs expressed by the earliest infiltrating cells, before a full-blown inflammatory response has occurred.

The Recruitment of a T-Cell Repertoire to Other Determinants on the
Same Antigen as well as on Other Antigens

After appropriate activation and expansion, autoantigen-specific T cells infiltrate the target tissue and result in local inflammation, cytokine release (e.g. IFN-γ) and tissue injury. As a result, processing and presentation of other tissue-specific antigens occurs, priming T cells with different reactivities. For example, it has been shown that MBP-primed diseased animals undergoing demyelination show responses to another CNS protein, PLP, as a result of in vivo priming [26, 27]. Similarly, responses also spread intramolecularly to other determinants of a given protein. Although the initial response to MBP was restricted to Ac1–9, the dominant determinant, later, several additional determinants of MBP recalled proliferative responses in the spleen of (B10.PL × SJL)F1 mice, namely those within peptides 35–47, 81–100 and 121–140 [28]. More pertinently, reactivity to the latter determinants was detected several weeks after induction of EAE with MBP peptide Ac1–11 alone. This demonstration indicated that the broadening of reactivity to additional MBP peptides involved endogenous MBP processing and presentation following priming of self Ac1–11-reactive T cells. It is not known what role these cells directed against the broader determinant repertoire play during the natural course of the disease. We found that except for 121–140, other cryptic determinanats on MBP are not able to induce EAE in B10.PL or (B10.PL × SJL)F1 mice [29;

Kumar, unpubl.]. This could be due to either a repertoire difference (Th1 vs. Th2) or some regulatory influences.

T Cells with Higher Affinity Receptors May Be Already Tolerized

Tolerance usually involves only those determinants on an autoantigen that are efficiently displayed to the T-cell pool (dominant self), sparing poorly presented determinants on the molecule (cryptic self). Adult mice, rendered tolerant to HEL by intravenous injection of the protein, do not respond to a subsequent challenge with the whole protein or peptides encompassing dominant determinants. However, such mice mount T-cell responses to cryptic determinants [30]. Supporting this view, Ria et al. [31] have shown that adult tolerance induction to the chimeric peptide formed by two peptides results in abolition of only the T-cell response against the dominant determinant. The same constellation has recently also been observed during spontaneous development of tolerance to endogenous antigens. Thus, H-2^d mice transgenic for HEL were nonresponsive to its single, dominant E^d-restricted determinant but did respond to cryptic determinants on HEL [32]. These data strongly suggest that self tolerance is only induced to efficiently presented, dominant determinants, but not to weakly presented, cryptic determinants on autoantigens.

We predict that dominant self peptides will have induced tolerance of all T cells with receptors of high affinity, that would have played a potentially dangerous role in autoimmune disease. However, many T cells able to respond at very low affinity to dominant self determinants should not be rendered tolerant. Either removal or regulation of high-affinity, self-directed T cells is undoubtedly one of the mechanisms important in maintaining tolerance to self. The remaining T cells directed against the cryptic self comprise both the autoreactive repertoire and the repertoire directed against foreign antigens.

Regulatory T Cells

An Equilibrium between Effectors and Regulators May Be Important in Autoimmunity

It has long been speculated that inappropriate immunoregulation plays a major role in the onset and maintenance of spontaneous autoimmune diseases such as IDDM. Various forms of T-cell depletion or inactivation induce the disease in healthy animals, suggesting regulatory mechanisms that control anti-islet autoimmune reactivity: (a) NOD mice either thymectomized at weaning [33] or treated with cyclophosphamide; (b) BB rats irradiated or treated with anti-RT6 monoclonal antibodies [34]; (c) NOD mice [35] and BB rats [36] protected from diabetes by injection of spleen cells from diabetes-free animals. These systems suggest a downregulatory cell type(s) active in maintaining homeostasis.

Although recognition of self determinants is essential for autoimmunity, it also can occur without autoimmune pathology. In rats, EAE could be prevented by treatment with anti-CD4; splenocytes from animals so treated were fully competent to transfer disease after in vitro activation [37]. Similarly, treatment with an anti-IgD-MBP peptide conjugate protects rats from EAE but allows the in vivo priming of encephalitogenic T cells [38].

If these immunoregulatory mechanisms malfunctioned, the resulting deficiency of regulatory T cells could lead to autoimmunity. For example, depletion of a particular subset of T cells results in thyroiditis in mice [39]; athymic rats reconstituted with CD45RB^{high} CD4+ T cells alone develop a severe autoimmune condition but when reconstituted with both CD45RB^{high} and regulatory CD45RB^{low} T cells they were devoid of autoimmune inflammation [40]. These and other neonatal depletion and reconstitution experiments suggest that a determining factor in the expression of autoimmunity is the equilibrium between autoreactive and regulatory T cells [41, 42].

TCR-Based Regulation of EAE

EAE is a T-cell-mediated demyelinating disease of the central nervous system and is generally self-limiting in B10.PL mice. A majority of MBP-induced T cells recognize N-terminal fragment Ac1–9, and predominantly utilize TCR Vβ8.2 and Vα2.3 variable (V) gene segments [22]. We have addressed the issue of whether the TCR itself can be processed and presented in the context of either class I or class II molecules, in vivo, in a physiological manner. If so, do T cells activated by TCR peptides become involved in the regulation of T-cell responses to MBP normally as well as during the course of the disease?

Experiments employing a T-cell vaccination approach by Cohen and co-workers [43] suggested that motifs on the T-cell receptors themselves may be engaged in regulating the cells on which they are displayed. Recently, it has been shown that immunization with particular V region peptides induces regulation and protects rats from EAE [44, 45]. To map out all the potential determinants within the V regions of one prototypic T-cell receptor (Vβ8.2 and Vα2.3) expressed on a majority of Ac1–9-reactive T cells in B10.PL mice [22], we have synthesized five 25–30-mer overlapping peptides from each TCR chain, encompassing the entire variable region. In testing their capacity to induce T-cell proliferation in B10.PL mice, we have found that three immunogenic self determinants are prevalent in the Vβ chain and two in the Vα chain. These determinants are spread along the V region and are not from the previously reported CDR II region that induced regulation of EAE in Lewis rats [46].

Thus, we found that three Vβ-chain peptides B2 (21–50), B4 (61–90) and B5 (76–101) were immunogenic when injected into B10.PL mice as peptides.

However, based on the proliferative recall responses after immunization with T cells expressing the receptor, there appears to be only a single immunodominant TCR determinant, B5, on the β-chain [46]. Thus, B5 appears dominant, while the determinants on B2 and B4 can be considered cryptic. Since, antigen processing and the structural context of determinants influence dominance, it is important to discuss some of the issues regarding the presentation of B5 in vivo. Importantly, we found that both cloned CD4+ T cells reactive to B5 as well as the peptide itself are capable of protecting mice from MBP-induced EAE, indicating an important role for presentation of B5 in the natural recovery from the disease.

Self-TCR peptide B5 can induce regulatory T cells that specifically downregulate Ac1–9 responses. When injected prior to MBP immunization, the immunodominant β-chain determinant B5 is specifically able to suppress the Ac1–9 proliferative response in both lymph nodes and spleens of B10.PL, PL/J and (SJL/J×B10.PL)F1 mice [46]. B5 immunization also significantly protects these mice from antigen-induced EAE [46]. Interestingly, other TCR determinants like B2 or B4, although strongly immunogenic, do not downregulate Ac1–9 responses, suggesting that two distinct functional sets of T cells are involved. As a control throughout these studies, we have examined the response to other determinants of guinea pig MBP (gpMBP), not present on mouse MBP (mMBP): responses to these unique gpMBP determinants, which presumably do not utilize Vβ8 genes, are not affected [14, 46].

We have generated T-cell lines and clones reactive to the immunodominant TCR-peptide B5. These T cells express the CD4 coreceptor and are restricted to the I-A^u class II molecule. Upon adoptive transfer in vivo in relatively small numbers (1×10^5 activated T cells), the TCR-peptide B5-specific T-cell clone B5.2 is able to downregulate Ac1–9 responses in recipient B10.PL mice and protect mice from antigen-induced EAE [14]. These regulatory T-cells (Treg) utilize a very limited set of TCR Vβ gene segments (Vβ14 and Vβ3), and a large majority of T cell clones utilize the Vβ14 gene segment [14].

Do mice recover from EAE because of a TCR-based T-T regulatory network which becomes operative during the course of the disease and suppresses the initial encephalitogenic response? To address this crucial question directly, we examined the B10.PL mouse after its initial episode of MBP-induced EAE and asked whether peripheral T cells specific for the immunodominant TCR β-chain peptide B5 had been induced via antigenic stimulation. Indeed, without any previous exogenous challenge with the TCR peptide, B5-specific Treg become primed in vivo during the initial challenge with antigen [14]. Preliminary experiments suggest that the deletion of Vβ14+ T cells from the peripheral T-cell repertoire indeed leads to severe EAE and poor recovery from EAE in (B10.PL × SJL)F1 mice [47]. These data further support our view that the Vβ14+ CD4 T cells constitute a crucial part of a regulatory T cell network responsible for recovery from EAE.

How do B5-specific regulatory T cells regulate responses to MBP? CD8 deletion experiments suggest that once activated, the regulatory CD4+ T cells recruit CD8+ T cells which ultimately downregulate encephalitogenic T cells [14]. Which are the structures that could serve as targets for regulation by CD8+ T cells? It is likely that determinants from nascent TCR chains, subject to degradation within the T cell, enter the class I pathway and are presented on the cell surface continually in an MHC class I context [48]. Recognition of MBP-specific effector cells by TCR-idiopeptide-specific CD8+ T cells may thus be one of the major pathways of regulation. It is most interesting that CD4+ T cells appear to be essential in the activation of the CD8+ T cells. It is likely that TCR-peptide-specific CD4+ T cells provide help for the priming of CD8+ T cells in vivo has been shown for a viral peptide recently [49]. One key prediction is that mice lacking regulatory CD8+ T cells should not be able to recover well from EAE and also should be susceptible to repeated relapses. Experiments in mice devoid of CD8+ T cells either by anti-CD8 treatment or CD8 gene targeting confirm this prediction [50, 51]. It is also likely that mice such as the SJL/J strain which readily gets chronic EAE have inefficient regulation.

It is likely that certain dominant TCR regulatory peptides are most readily presented to T cells following antigen processing, as is the pattern in class II responses against conventional antigens. It could also be true that the favored determinant lies within a framework region appearing in different V gene families, which would give rise to very broad regulatory families. There could be a hierarchy of broadly public as well as very private TCR idiopeptides that would permit both types of regulatory cells to gain access to a single target T-cell. Accordingly, a network of V domains could be structured in which many of the cellular participants would be regulatory T cells.

It seems unlikely that evolution would have favored an immune system spending a large portion of its resources on specific regulation. Therefore, a focus on particular dominant public motifs of certain V regions of the TCR can provide coordinate and effective regulation, economically. It is possible that these dominant TCR motifs are presented by activated antigen-specific B-cells, although TCR-idiotope-specific B cells must first be activated. Interestingly, significant levels of anti-TCR antibodies are induced in animals recovering from MBP or Ac1–9-induced EAE [Kumar, unpubl.]. Activation of such anti-TCR specific B cells may be an important factor in recruiting a cascade of regulatory players in vivo. Therefore, in the mouse system, processing and presentation of TCR peptides by professional APCs may be a determining event in the induction of TCR-based regulation.

Human T cells expressing class II MHC molecules have been shown to efficiently process and present antigen [52]. Recent evidence shows that mouse T cells also display class II MHC molecules as well as B7, temporarily upon activation, and they should therefore be able to present self TCR peptides. Otherwise, it is possible that the TCR-specific B cells mentioned previously could pick up TCRs at the inflammatory site and present TCR peptides in the context of class II MHC molecules to CD4+ regulatory T cells. Since B cells which express a specific antibody are highly efficient APC for that antigen, capable of activating T cells when provided with 1/1,000th the antigen concentration required for non-specific APC [53, 54], a minimal number of TCR molecules should be sufficient to turn on regulatory T cells in vivo. Another possibility is that dendritic cells, which are very efficient APC may be involved in the processing and presentation of TCR peptides.

Interestingly, we have found that spleen cells from naive animals are capable of stimulating TCR-peptide-reactive T cells in vitro in the absence of exogenous peptide [14]. There are two explanations for this observation. (i) APCs constitutively process and potentially present B5 in vivo. In view of the prevalence of the TCR Vβ8 family in peripheral T cells (15–30%), one may imagine that TCR Vβ8 peptides could be generated that are capable of binding to MHC molecules in vivo. TCR-specific B cells would be the most parsimonious solution to this requirement for presentation, since they require a minimal quantity of specific antigen. (ii) Another possible interpretation is that some other molecule(s) on the surface of splenic APC could be cross-reactive [55] with the B5 determinant. We are currently characterizing the peptide isolated from these APC.

Typically, dominant determinants on self-antigens have induced tolerance [30]. Thus, why is the T-cell repertoire directed against dominant TCR peptide B5 still intact! Are these TCR-peptide/MHC complexes not present in the thymus or do they fail to mediate negative selection? Preliminary data suggest that the TCR-peptide B5-specific T-cell repertoire is not tolerized even after neonatal treatment of mice with B5, unlike the situation with all other peptides we have used. It is tempting to speculate that T-cell regulatory networks may be operative through the recognition of some dominant TCR-peptides, for example, B5 [46, 56], that are part of a 'homuncular network' [11]. One of the necessities in such a network is that regulatory cells not themselves be susceptible to removal through tolerance. Therefore, understanding the processing and the presentation of the self-TCR peptides in vivo is extremely important to explore the mechanistic rules governing TCR-based network regulation.

Interestingly, B5-reactive T cells are not revealed before the onset of EAE in mice immunized with MBP/CFA and pertussis toxin (PT)! There are at least two explanations for this phenomenon; (i) Regulatory TCR Vβ14+ T cells are only primed after the expansion of MBP-reactive Vβ8.2+ T cells which induce EAE, and subsequently only upon optimum priming and expansion of B5-reactive T cells which results in the down-regulation of encephalitogenic Ac1–9-reactive T cells. (ii) Immunization with MBP in CFA and pertussis toxin somehow inhibits the normal initial expansion of Vβ14+ T cells temporarily; upon release of inhibition, these cells expand leading to the downregulation of the Vβ8.2+ effector T cells. Preliminary data suggest that B5-reactive T cells appear to be present normally but in an inhibited state. In conclusion, it is likely that the TCR Vβ14+ regulatory T cells are an essential part of a physiological T-cell regulatory circuit that maintains and reestablishes tolerance to self MBP. The organism can withstand the perturbations and flare-ups that probably characterize the delicate balance inherent in the immune system. It appears that disruption of this cellular circuitry can lead to a failure to maintain an overt equilibrium, resulting in autoimmunity.

Antigen-Centered Regulation

In two antigen systems, hen eggwhite lysozyme (HEL) and *Escherichia coli* β-galactosidase (GZ), we have demonstrated the existence of suppressive T cells which were activated by particular peptides from the antigen and not others, so-called 'suppressor determinants = SD' [57–59]. In each case, CD8+ T cells were activated, which were able to negate the activitiy of CD4+ helper T-cells, directed against other determinants (HD) on the antigen.

One instructive relationship between these two cell types, pointing out the interaction of antigen processing and suppressive cells in determining an eventual response phenotype, has been clarified recently in the HEL system. We have known for many years that several H-2^b strains, e.g. C57BL/6 were responders, and other H-2^b, e.g. C3H.SW or BALB.B, were nonresponders to HEL [60]. Likewise, earlier work had shown that suppressor T cells were present in each of the above strains, and that these Ts appeared in the spleens but not the lymph nodes of mice 10 days after HEL-CFA priming. Thus, even in the 'nonresponsive' strains, a proliferation response was evident in the draining popliteal lymph nodes, but not in the spleen.

When the specificity of this proliferative response was tested in spleen and lymph nodes of each strain, responders demonstrated a response to 46–61 in the spleen and a response to 20–35, 30–53, 46–61, and 74–96 in the lymph node. Meanwhile, nonresponders gave a response in the lymph nodes to 20–35, 30–53 and 74–96, but not to 46–61, and there was no response whatever in the spleen to foodpad or to intraperitoneal injection of HEL. We could conclude from these data that the Ts in the spleen of both strains were able to prevent responses in T cells specific for 20–35, 30–53 and 74–96, but that in the responder strains, the Ts were unable to regulate the response to 46–61.

But why was there a response to 46–61 in the C3H.SW but not in the C57BL/6 mouse? Grewal et al. [60] recently showed that 46–61 could not bind to the I-A^b molecule whereas 46–60 bound very efficiently. The responsive C3H.SW must first process 46–61 into 46–60 before binding occurs. Thus, the C57BL/6 non-MHC genes must be defective in their ability to remove the C-terminal arginine 61 from the precursor of the binding peptide, presumably lacking a carboxypeptidase. This concentration of events in the C57BL/6 animal seems to be responsible for the total lack of T-cell responsiveness to HEL in this animal's spleen, which earlier was also assayed as a lack of antibody-forming cell capacity, and therefore classified as Ir-gene unresponsiveness.

Genetic Unresponsiveness

Genetic unresponsiveness to protein antigens has generally been attributed to three different mechanisms: the failure to bind to MHC molecules [61]; or to 'holes in the T-cell repertoire', even when the peptide binds to MHC [62, 63]; or to regulatory effects [64]. The absence of an appropriate T-cell repertoire capable of recognizing the antigen has been attributed primarily to clonal deletion during thymic tolerance induction [65] or to peripheral clonal exhaustion [3, 4] after a vigorous response.

Lack of Binding to the MHC

Binding of determinants to MHC class II molecules is a competitive act, with other determinants both up and downstream on the unfolding antigen competing for primacy in binding to the class II molecules in the appropriate acidic compartment of the cell. Usually, as a function of the degree of availability of the agretope and the relative binding energies of the different agretopes on the competing determinants, one or a very few determinants will emerge as winners in the competition [66]. It is a rare event that not even one binding determinant will be present on a molecule of 100 amino acids or more, and in fact, not a single such situation has been reported. Therefore, this explanation for genetically determined Ir gene unresponsiveness is unlikely, although if one reduces the scope of the question to the binding of single determinants, cases are known and described below. For example, in one instance, a strongly binding determinant exists, but this binding is prevented if there is a flanking amino acid of the wrong type present [60].

Processing Defect May Lead to Non-Responsiveness

Several factors determine the level of binding of a given peptide to the MHC, including its availability following antigen processing. A processing defect resulting in genetic unresponsiveness to a hen egg lysozyme determinant by failure to remove a single amino acid which hinders peptide binding to the MHC has recently been shown in our laboratory [60]. Such a 'hinderotope' may interfere with response to a particular T-cell determinant. Similarly, specific requirements for amino acid residues with either positive charge or hydrophobicity, for effective loading by peptide transporters, may also limit the spectrum of potential determinants on an autoantigen [67].

Absence of Critical Residues within CDR3 Constrains T-Cell Recognition of a Self Antigen

T cells derived from B10.PL or PL/J mice of the H-2^u haplotype, utilize only Dβ2 and Jβ2 gene segments in the recognition of the dominant determinant, Ac1–9/A^u, of myelin basic protein (MBP). NZW mice, with identical class II H-2^u genes (I-A and I-E), carry an 8.8 kb deletion in their TCR β-chain locus encompassing Dβ2 and Jβ2 gene segments [68]. How does this deletion of the crucial Dβ2-Jβ2 region in NZW mice influence specific responses to Ac1–9/A^u as well as to other known A^u or E^u determinants of MBP? We found that these mice respond very poorly to the dominant Ac1–9/A^u and to the subdominant 31–50/E^u determinant in T-cell proliferation assays or in the induction of experimental autoimmune encephalomyelitis (EAE) [63]. This loss of response is apparently owing to the absence of high avidity TCRs with essential CDR3 residues contributed by Dβ2 or Jβ2 gene segments. These data demonstrate constraints in the recognition of certain antigenic structures in spite of the myriad possibilities for TCR diversity. Presumably, the lack of particular dominant T cells with high avidity TCRs results in functional nonresponsiveness even to strong antigenic determinants.

Although the presence of extremely low-affinity/frequency T cells of Ac1–9/A^u or 35–47/E^u specificity in NZW mice cannot be ruled out, their activity is not detectable with in vitro or in vivo assays. It is possible that low-affinity T cells specific for these determinants still exist and are of the Th2 type. These cells usually do not proliferate well and are likely to be poorly encephalitogenic. Thus, it is likely that holes in the repertoire do exist, probably due to the absence of TCRs employing appropriate Dβ-Jβ gene segments. It seems that the interaction between the TCR structures and their ligands, Ac1–9/A^u or 31–50/E^u complexes, is stringent enough to demand not only specific Vβ gene segments but also specific residues in the CDR3 region (Dβ2-Jβ2), despite the availability of closely related gene segments within the Dβ1-Jβ1 region [63].

Susceptibility to MBP-Induced EAE May Correlate with the Balance between Th1 and Th2 Cells

The induction of EAE is associated both with particular MHC and non-MHC alleles. Why does MBP immunization not lead to EAE in some of the susceptible MHC haplotypes? SJL/J mice are susceptible while A. SW are not susceptible to MBP-induced EAE, although both of these strains express the susceptible MHC haplotype, H-2^s. We have studied the lymph node T-cell-proliferative responses to MBP in these and other H-2^s mice and found that non-disease-susceptible strains appear to be poor responders. It is interesting that although A. SW mice respond poorly to whole MBP, the central immunodominant determinant of MBP in the H-2^s haplotype (MBP 81–100) induces very efficient proliferative responses as well as EAE [Kumar, unpubl.]. Our data suggest that the poor responses to MBP may correlate with either the absence of Igh-b or the presence of the Igh-e locus in some of the congenic strains that we have tested [Kumar, unpubl.]. At present, we do not know how the Ig locus affects the susceptibility to EAE. It is possible that the Igh locus may affect the developing T-cell repertoire through idiotypic interaction and may lead to the skewing of a subpopulation of MBP-reactive cells towards the Th2 phenotype. This becomes more relevant in the context of the demonstration that different mouse strains can respond to the same peptide and give rise to clones that produce either IL-2 or IL-4 [69]. Therefore, if EAE were only induced by γ-IFN-producing Th1 clones, it is plausible that in a nonsusceptible strain, IL-4 producers might preferentially expand and indirectly protect the animals from EAE. Preliminary data suggest that in mice with the Igh^b locus, the initial antigen-specific response is dominated by Th2-like cells but not in mice with the Igh^e locus [Kumar, unpubl.].

Th1 and Th2 Balance May Be Crucial for Autoimmunity

It has been clearly demonstrated that self tolerance to at least some tissue specific antigens is not entirely a passive process but rather an active dynamic state in which potentially pathogenic self-reactive T cells are prevented from causing disease by other regulatory T cells [69]. The differential secretion of lymphokines by T-cell subsets offers an explanation for the ability of certain T cells to induce autoimmunity and others to regulate these autoreactive T-cells [70].

A functional heterogeneity of CD4+ T cells based on the expression of different isoforms of CD45 has been shown in both rats and recently in mice [40, 71]. There appears to be an association between the CD45 phenotype on cells and their potential to secrete certain cytokines [71]. CD45RC^{high} cells produce IFN-γ and IL-2 and induce graft vs. host as well as wasting disease. In contrast, CD45RC^{low} cells seem regulatory and suppress pathology while secreting high levels of IL-4. This subset has been shown to inhibit diabetes in BB rats.

Th1 cytokines have been shown to be involved in cell-mediated autoimmune diseases. For example, IFN-γ, lymphotoxin or TNF-α secretion correlate with the encephalitogenic capacity of T-cell clones reactive to MBP [72]. Anti-TNF-α antibodies have been shown to inhibit EAE and collagen-induced arthritis while anti-TGF-β accelerates the disease [70, 71]. It seems that both TGF-β and IL-10 play an important role in regulating autoimmune inflammatory reactions. Knockout mice for these genes develop a chronic inflammatory condition involving various organs [73, 74]. However, it is important to emphasize that generalized effects of different cytokines are unpredictable, owing to the temporal and positional induction of cytokines during an immune response.

Both Th1 and Th2 cells differentiate from a common pool of precursors postthymically and as a consequence of activation by the antigen [75]. Cytokines produced by Th1 and Th2 cells play an important role in mutual regulation of their differentiation and subsequent effector functions [76, 77]. Do interactions among the MHC/peptide, TCR and other accessory or costimulatory molecules affect lymphokine secretion as well as the differentiation of Th cells? An understanding of the involvement of these interactions in the differential cytokine secretion pattern by T cells of a given specificity is crucial.

We asked whether ligands with high or low MHC-binding capabilities could result in tilting the response towards Th1 or Th2. The N-terminal nonapeptide of MBP was chosen as the test system: It has been shown that amino acid 4 is involved in MHC binding and by altering this position 4 from Lys to Met, for example, the affinity of MHC interaction increases 10^4-fold [78–80]. B10.PL mice were immunized with the position 4 variant nonapeptides (with 4Arg, a 3-fold poorer binder than 4Lys) and 10 days later, frequencies of Th1 or Th2 cells were assessed in draining lymph nodes using cytokine-specific, single-cell, enzyme-linked immunospot (ELISA SPOT) assays. The frequency of IFN-γ-producing cells was about 5 times higher in mice challenged with the highest affinity peptide, 4Met, in comparison with mice immunized with one of the low-affinity peptides. In contrast, the frequency of IL-5-producing cells was comparable in all three groups of mice. Thus, the high-affinity ligand appeared to shift the lymphokine-secretion pattern more towards the Th1 type. Consistent with the involvement of Th1 cells in EAE, in B10.PL mice the highest affinity ligand was able to effectively induce disease at very low concentrations [Kumar, unpubl.].

Although all variants are able to induce proliferation and IL-4 secretion, only the variant with 4Met induced significant levels of IFN-γ secretion at a relatively high peptide concentration. These data are consistent with the idea that changes in the binding affinity of a ligand to the MHC result in altered signalling owing to differences in either the display of the ligand (MHC-peptide complexes) or in the quality of the ligand on the surface of APC [78, 81]. According to this model, interaction of the TCR with ligands displayed either sparsely or densely on the surface of APC will determine the cytokine secretion profile of the T cell. Alternatively, the high-affinity peptide form relatively a stable complex with the MHC moleculs resulting in high avidity interaction with the same TCR and leading to Th1 induction. These data point to a flexibility of T-cell effector function that is very dependent on the physical nature of the tripartite interactions among the MHC, antigen and the TCR.

Recent demonstrations regarding binding of the TCR to altered peptides resulting in differential signalling [82] are consistent with the idea that the nature of the entire stimulus brought about by various interactions, including co-receptor (e.g. CD4) and accessory (e.g. CD28) molecules, could elicit different T-cell effectors with a unique pattern of cytokine secretion. Whatever may be the mechanism of skewing response towards Th1 or Th2 that results from altering the physical interaction between the TCR and the ligand, these data offer new avenues for therapy in diseases, for example, autoimmunity, allergy or AIDS, where the T-cell response must be redirected into a healthy pattern.

References

1   Schwartz R: A cell culture model for T lymphocyte clonal anergy. Science 1990;248:1349–1356.
2   Miller JFAP, Morahan G: Peripheral T cell tolerance. Annu Rev Immunol 1992;10:51–69.
3   Webb S, Morris C, Sprent J: Extrathymic tolerance of mature T cells. Clonal elimination as a consequence of immunity. Cell 1990;63:1249–1256.
4   Byers VS, Sercarz EE: The X-Y-Z scheme of immunocyte maturation IV. The exhaustion of memory cells. J Exp Med 1968;127:307–325.
5   Stockinger B, Lin RH: An intercellular self-protein synthesized in macrophages is presented but fails to induce tolerance. Int Immunol 1989;1:592–596.
6   Eynon EE, Parker DC: Do small B cells induce tolerance. Transplant Proc 1991;23:729–730.
7   Kumar V, Sercarz E: The involvement of TCR-peptide-specific regulatory CD4+ T cells in recovery from antigen-induced autoimmune disease. J Exp Med 1993;178:909–916.
8   Kumar V, Sercarz E: T cell regulatory circuitry: Antigen-specific and TCR-idiopeptide-specific T cell interactions in EAE. Int Rev Immunol 1992;9:269–279.
9   Karpus WJ, Gould KE, Swanborg RH: CD4+ suppressor cells of autoimmune encephalomyelitis respond to T cell receptor associated determinants on effector cells by IL-4 secretion. Eur J Immunol 1992;22:1757.
10 Myers L, Stuart J, Kang A: A CD4 cell is capable of transferring suppression of collagen-induced arthritis. J Immunol 1989;143:3976–3980.

11 Cohen IR, Young DB: Autoimmunity, microbial immunity and the immunological homunculus. Immunol Today 1991;12:105–110.

12 Heber-Katz E, Acha-Orbea H: The V-region hypothesis: Evidence from autoimmune encephalomyelitis. Immunol Today 1989;10:164.

13 Sercarz E, Araneo B, Benjamin CD, Harvey M, Metzger D, Miller A, Wicker L, Yowell R: The design of regulatory circuitry: predominant idiotypy and the idea of regulatory parsimony. Ann NY Acad Sci Immune Networks 1983;418:198–205.

14 Kumar V, Sercarz E: Regulation of autoimmunity. Curr Opin Immunol 1991;3:888–895.

15 Mosmann TR, Coffman RL: Th1 and Th2 cells: Different pattern of lymphokine secretion lead to different functional properties. Ann Rev Immunol 1989;7:145–173.

16 Eichmann K, Falk I, Rajewsky K: Recognition of idiotype interactions. Eur J Immunol 1978;8:853–857.

17 Elias D, Markovitz D, Reshef T, van Der Zee R, Cohen I: Induction and therapy of autoimmune diabetes in NOD mouse by a 65 kDa heat shock protein. Proc Natl Acad Sci USA 1990; 87:1576–1580.

18 Elias D, Reshef T, Birk OS, van Der Zee R, Walker MD, Cohen IR: Vaccination against autoimmune mouse diabetes with a T cell epitope of the human 65 kDA heat shock protein. Proc Natl Acad Sci USA 1991;88:3088–3091.

19 Kaufman DL, Clare-Salzer M, Tian J, Forsthuber T, Ting GSP, Robinson P, Atkinson MA, Sercarz E, Tobin AJ, Lehmann PV: Spontaneous loss of T cell self tolerance to glutamate decarboxylase is a key event in the pathogenesis of murine insulin-dependent diabetes. Nature 1993; 366:69–72.

20 Kumar V, Kono DH, Urban JL, Hood L: T-cell receptor repertoire and autoimmune diseases. Ann Rev Immunol 1989;7:657–682.

21 Acha-Orbea H, Steinman L, McDevitt HO: T cell receptors in murine autoimmune diseases. Ann Rev Immunol 1989;7:371–405.

22 Urban JL, Kumar V, Kono DH, Gomez C, Horvath SJ, Clayton J, Ando DG, Sercarz EE, Hood L: Restricted use of T cell receptor V genes in murine autoimmune encephalomyelitis raises possibilities for antibody therapy. Cell 1988;54:577–592.

23 Sellins KS, Danska JS, Paragas V, Fathman CG: Limited TCR β-chain usage in the sperm whale myoglobin 110–121 response by H-2^d congeneic mouse strains J Immunol 1992;149:2323.

24 Osman GE, Toda M, Kanagawa O, Hood L: Characterization of the T cell receptor repertoire causing collagen arthritis in mice. J Exp Med 1993;177:387–395.

25 Candeias S, Katz J, Benoist C, Mathis D, Haskins K: Islet-specific T cell clones from NOD mice express heterogeneous T cell receptors. Proc Natl Acad Sci USA 1991;88:6167–6170.

26 McCarron RM, Fallis RF, McFarlin DE: Alterations in T cell antigen specificity and class II restriction during the course of chronic relapsing experimental allergic encephalomyelitis. J Neuroimmunol 1990;29:73–79.

27 Perry LL, Barzaga-Gilbert E, Trotter JL: T cell sensitization to proteolipid protein in myelin basic protein-induced relapsing experimental allergic encephalomyelitis. J Neurimmunol 1991; 33:7–15.

28 Lehmann PV, Forsthuber T, Miller A, Sercarz EE: Determinant spreading in autoimmunity: cryptic T-cell determinants of myelin basic protein become immunogenic after immunization with a dominant peptide of the molecule. Nature 1992;366:69–72.

29 Bhardwaj V, Kumar V, Grewal I, Dao T, Lehmann P, Geysen M, Sercarz E: The T cell determinant structure of myelin basic protein in B10.PL, SJL, and their F1s. J Immunol 1994;152: 3711–3719.

30 Gammon G, Sercarz E: How some T cells escape tolerance induction. Nature 1989;342: 183–185.

31 Ria F, Chan BMC, Scherer MT, Smith JA, Gefter ML: Immunological activity of covalently linked T cell epitopes. Nature 1990;343:381–383.

32 Cibotti R, Kanellopoulos J, Cabaniols JP, Halle-Panenko O, Kosmatopoulos K, Sercarz E, Kourilsky P: Tolerance to a self-protein involves its immunodominant but not subdominant determinants. Proc Natl Acad Sci USA 1992;89:416–420.

33   Dardenne M, Lepault F, Bendelac A, Bach JF: Acceleration of the onset of diabetes in NOD mice by thymectomy at weaning. Eur J Immunol 1989;19:889–895.

34   Greiner DL, Mordes JP, Handler ES, Angelillo M, Nakamura N, Rossini AA: Depletion of RT 6.1+ T lymphocytes induces diabetes in resistant biobreeding/worcester (BB/W) rats. J Exp Med 1987;166:461–475.

35   Boitard C, Yasunamy R, Dardenne M, Bach JF: T cell-mediated inhibition of the transfer of autoimmune diabetes in NOD mice. J Exp Med 1989;169:1669–1680.

36   Rossini AA, Faustman D, Woda BA, Like AA, Szymanski I, Mordes JP: Lymphocyte transfusions prevent diabetes in the bio-breeding/worcester rat. J Clin Invest 1984;74:39–46.

37   Sedwick JD, Mason DW: The mechanism of inhibition of experimental allergic encephalomyelitis in the rat by monoclonal antibody against CD4. J Neuroimmunol 1986;13:217–232.

38   Day MJ, Tse AGD, Puklavic M, Simmonds SJ, Mason DW: Targetting autoantigen to B cells prevents the induction of a cell mediated autodisease in rats. J Exp Med 1992;175:655–659.

39   Sugihara S, Maruo S, Tsujimura T, Tarutani O, Kohno Y, Hamaoka T, Fujiwara H: Autoimmune thyroiditis induced in mice depleted of particular T cell subset: III. analysis of regulatory cells suppressing the induction of thyroiditis. Int Immunol 1990;2:343–351.

40   Powrie F, Mason D: OX-22^{high} CD4+ T cells induce wasting disease with multiple organ pathology: prevention by the OX-22^{low} subset. J Exp Med 1990;172:1701–1708.

41   Kumar V, Tabibiazar R, Sercarz E: The maintenance of peripheral tolerance through TCR-based regulation. Autoimmunity 1993;14(suppl)2.

42   Kumar V, Sercarz E: In Zouali M (ed): Maintenance and Re-Establishment of Self-Tolerance: TCR-Peptide-Specific Regulatory T Cells. Autoimmunity: Experimental Aspects. Berlin, Springer, 1994, pp 29–38.

43   Ben-Nun A, Wekerle H, Cohen IR: Vaccination against autoimmune encephalomyelitis using attenuated cells of a T lymphocate line reactive against myelin basic protein. Nature 1981;292:60–61.

44   Howell MD, Winters ST, Olee T, Powell HC, Carlo DJ, Brostoff SW: Vaccination against experimental allergic encephalomyelitis with T cell receptor peptides. Science 1989;246:668–670.

45   Vandenbark AA, Hashim G, Offner H: Immunization with a synthetic T-cell receptor V-region peptide protects against experimental autoimmune encephalomyelitis. Nature 1989;341:541–544.

46   Kumar V, Tabibiazar R, Geysen M, Sercarz E: The dominant self-TCR peptide from the framework III region of the β-chain protects mice from EAE. In preparation.

47   Kumar V, Stellrecht K, Meyer M, Sercarz E: Deleting CD4 cells from a TCR-based cellular circuitry controlling EAE leads to chronic disease (abstract). J Immunol 1994;152:3229.

48   Jiang H, Zhang S, Pernis B: Role of CD8+ T cells in experimental allergic encephalomyelitis. Science 1992;256:1213–1215.

49   Fayolle C, Deriaud E, Leclerc C: In vivo induction of cytotoxic T cell response by a free synthetic peptide requires CD4+ T cell help. J Immunol 1991;147:4069–4073.

50   Jiang H, Sercarz E, Nitecki D, Pernis B: The problem of presentation of T cell receptor peptides by activated T cells. Ann NY Acad Sci 1991;636:28–32.

51   Koh D-R, Fung-Leung W-P, Ho A, Gray D, Acha-Orbea H, Mak T-W: Less mortality but more relapses in experimental allergic encephalomyelitis in CD8–/– mice. Science 1992;256:1210–1213.

52   Lanzavecchia A, Roosnek E, Gregory T, Berman P, Abrignani S: T cells can present antigens such as HIV gp120 targeted to their own surface molecules. Nature 1988;334:530–532.

53   Casten LA, Kaumaya P, Pierce SK: Enhanced T cell responses to antigenic peptides targeted to B cell surface Ig, Ia, or class I molecules. J Exp Med 1988;168:171–180.

54   Lanzavecchia A: Antigen-specific interactions between T and B cells. Nature 1985;314:537–539.

55   Bhardwaj V, Kumar V, Geysen HM, Sercarz E: Degenerate recognition of a dissimilar antigenic peptide by MBP-reactive T cells: Implications for thymic education and autoimmunity. J Immunol 1993;151:5000–5010.

56   Kumar V, Tabibiazar R, Sercarz E: A homuncular T cell network exists in the B10.PL mouse controlling autoimmunity and based on TCR-peptide recognition by regulatory T cells. J Immunol 1993;150:174A.

57 Asano Y, Hodes RJ: T cell regulation of B cell activation, an antigen mediated tripeptide interaction of Ts cells, Th cells and B cells is required for suppression. J Immunol 1984;133: 2864–2867.

58 Adorini L, Harvey MA, Miller L, Sercarz EE: The fine specificity of regulatory T cells. II. Suppressor and helper T cells are induced by different regions of HEL in a genetically non-responder mouse strain. J Exp Med 1979;150:293–306.

59 Sercarz E, Krzych U: The distinctive specificity of antigen-specific suppressor T cells. Immunol Today 1991;12:111–118.

60 Grewal I, Moudgil K, Sercarz EE: An antigen processing defect underlies genetic unresponsiveness by failure to remove a single amino acid hindering peptide binding to MHC. 1994; submitted.

61 Buus S, Sette A, Colon SM, Miles C, Grey HM: The relation between major histocompatibility complex (MHC) restriction and the capacity of Ia to bind immunogenic peptides. Science 1987;235:1353.

62 Dos Reis GA, Shevach EM: Antigen-presenting cells from nonresponder strain 2 guinea pig are fully competent to present bovine insulin B chain to responder strain 13 T cells. Evidence against a determinant selection model and in favor of a clonal deletion model of immune response gene function. J Exp Med 1983;157:1287.

63 Kumar V, Sercarz E: Holes in the T cell repertoire to MBP owing to the absence of the Dβ2-Jβ2 gene cluster: Implications for TCR recognition and autoimmunity. J Exp Med 1994;179: 1637–1643.

64 Jensen PE, Kapp JA, Pierce CW: The role of suppressor T cells in the expression of immune response gene function. J Mol Cell Immunol 1987;3:267.

65 Kappler J, Roehm N, Marrack P: T cell tolerance by clonal elimination in the thymus. Cell 1987;49:273–280.

66 Sercarz E, Wilburx S, Sadegh-Nasseri S, Miller A, Manca F, Gammon G, Shastri N: The molecular context of a determinant influences its dominant expression in a T cell response hierarchy through fine processing. Prog Immunol 1986;6:227–237.

67 Momburg F, Roelse J, Howard JC, Butcher GW, Hammerling GJ, Neefjes JJ: Selectivity of MHC-encoded peptide transposons from human, mouse and rat. Nature 1994;367:648–651.

68 Kotzin BL, Barr VL, Palmer E: A large deletion within the T-cell receptor β chain gene complex in New Zealand white mice. Science 1985;229:167.

69 Pfeiffer C, Murray J, Madri J, Bottomly K: Selective activation of Th1- and Th2-like cells in vivo- Response to human collagen IV. Immunol Rev 1991;123:65–84.

70 Mason D, Fowell D: T-cell subsets in autoimmunity. Curr Opin Immunol 1992;4:728–732.

71 O'Garra A, Murphy K: T-cell subsets in autoimmunity. Curr Opin Immunol 1993;5:880–886.

72 Powell MB, Mitchell D, Lederman J, Buckmeier J, Zamvil SS, Graham M, Ruddle NH, Steinman L: Lymphotoxin and tumor necrosis factor α production by MBP specific T cell clones correlate with encephalitogenicity. Int Immunol 1990;2:539–544.

73 Shull MM, Ormsby I, Kier AB, Pawlowski S, Diebold RJ, Yin M, Allen R, Sidman C, Proetzel G, Calvin D, et al: Targeted disruption of the mouse transforming growth factor β-1 gene results in multifocal inflammatory disease. Nature 1992;359:693–699.

74 Kuhn R, Lohler J, Rennick D, Rajewsky K, Muller W: Interleukin-10 deficient mice develop chronic enterocolitis. Cell 1993;75:263–274.

75 Rocken M, Saurat JH, Hauser C: A common precursor for CD4+ T cells producing IL-2 or IL-4. J Immunol 1992;148:1031–1036.

76 Seder RA, Gazzinelli R, Sher A, Paul WE: IL-12 acts directly on CD4+ T cells to enhance priming of IFN-γ production and diminishes IL-4 inhibition of such priming. Proc Natl Acad Sci USA 1993;90:10188–10192.

77 Hsieh CS, Heimberger AB, Gold JS, O'Garra A, Murphy KM: Differential regulation of T helper phenotype development by IL-4 and IL-10 in an αβ transgenic system. Proc Natl Acad Sci USA 1992;89:6065–6069.

78 Kumar V, Bhardwaj V, Soares L, Sette A, Sercarz E: MHC-ligand affinity determines the productions of Th1- or Th2-like lymphokines by a T cell clone. J Cell Biochem 1994;(suppl 18D): 357,V379.

79 Kumar V, Bhardwaj V, Soares L, Alexander J, Sette A, Sercarz E: MHC-ligand affinity of an antigenic determinant is crucial for differential secretion of IFN-γ or IL-4 by T cells. Submitted.

80 Wraith DC, Smilek DE, Mitchell DJ, Steinman L, McDevitt HO: Antigen recognition in auto-immune encephalomyelitis and the potential for peptide-mediated immunotherapy. Cell 1989; 59:247–255.

81 Tao X, Pfeiffer C, Sette A, Bottomly K: Effect of peptide binding affinity on the development of CD4+ T lymphocyte effector function. J Cell Biochem 1994;(suppl 18D):362,V398.

82 Evavold BD, Sloan-Lancaster J, Allen PM: Tickling the TCR: Selective T-cell functions stimulated by altered peptide ligands. Immunol Today 1993;14:602–609.

Vipin Kumar, PhD, Department of Microbiology and Molecular Genetics,
University of California, Los Angeles, CA 90024–1489 (USA)

Adorini L (ed): Selective Immunosuppression: Basic Concepts and Clinical Applications.
Chem Immunol. Basel, Karger, 1995, vol 60, pp 20–31

....

Selective Targets for Immunotherapy in Autoimmune Disease

Xiao-Dong Yang[a], *Roland Tisch*[a], *Hugh O. McDevitt*[a, b]

Departments of [a]Microbiology and Immunology, and [b]Medicine,
Stanford University School of Medicine, Stanford, Calif., USA

Autoimmunity is due to the recognition and destruction of self components by the immune system, which normally functions to prevent invasion of foreign organisms. It is evident that autoimmunity is due to the failure of normal mechanisms of self tolerance [1, 2]. Therefore, the development of immunotherapy regimens relies heavily on our understanding of the mechanisms by which the immune system recognizes self, and how these recognition events are regulated. In this article, we will briefly note some current hypotheses regarding self tolerance and autoimmunity, in addition to discussing several novel approaches developed recently in our and other laboratories for the treatment of autoimmune diseases in animal models.

T-Cell Tolerance

The ability of the immune system to discriminate between self and foreign antigen has evolved to ensure that lymphocytes are tolerant to self while capable of responding to any foreign antigen [3]. Self tolerance is or can be established and maintained at different developmental stages, within different immune organs and probably via various different mechanisms.

Central Tolerance

Depending upon the organs where lymphocytes develop, tolerance can be classified as central and peripheral. Central tolerance mainly occurs within the thymus where immature T cells expressing self-reactive T-cell receptors (TCR) are deleted upon engagement with self antigen and major histocompatibility complex (MHC) molecules, a mechanism known as negative selection [4–6]. Thymic negative selection, however, is not complete since self-reactive lymphocytes can be detected in normal individuals, suggesting that autoreactive lymphocytes may be inactivated or controlled by other factors in the periphery, a mechanism known as peripheral tolerance [7].

Peripheral Tolerance

Peripheral tolerance can be achieved by various mechanisms including clonal anergy, peripheral deletion, ignorance and immune regulation and/or suppression.

Anergy. Some immunocompetent T cells which have escaped thymic negative selection and have emigrated to the periphery are found in an unresponsive state upon encountering specific antigen. There are at least two pathways to this state of unresponsiveness. First, it has been shown in a transgenic model that T-cell unresponsiveness can be due to a marked downregulation of the TCR, the TCR coreceptor CD8 (or CD4) or both. The second mechanism termed clonal anergy is an unresponsive state of T cells upon stimulation by specific antigen [8, 9]. T-cell activation requires not only the signal transduced by the TCR upon engagement with specific antigen/MHC, but also other costimulatory signals provided by interactions of CD4 or CD8 molecules on T cells with class II or class I MHC molecules on antigen-presenting cells (APC); the ligation of cell adhesion molecules with their ligands; or the action of cytokines. One important costimulatory pathway is the interaction of CD28 or CTLA-4 on T cells with B7 antigens expressed on APC [10–12]. Engagement of the TCR with MHC complexed with peptide without costimulatory signals may deliver a negative signal to T cells, leading to anergy. It has been shown that anergy can be overriden by exogeneous cytokines such IL-2, which may substitute as a costimulatory molecule, or induce the expression of costimulatory molecules.

Peripheral Deletion. The deletion of autoreactive lymphocytes has been considered to be a function of the thymus. Recent studies have clearly demonstrated, however, that in the periphery mature lymphocytes including autoreactive T cells can also be deleted. For example, it has been shown that mature T cells can be deleted in the periphery by administration of large doses of self antigen, superantigen or foreign antigens [13–17]. However, it is still unclear whether such mechanisms occur in normal, unmanipulated animals in order to maintain self tolerance. Furthermore, it remains unknown what the underlying mechanism(s) leading to peripheral deletion is. It appears that persistent stimulation with antigen while the T cell is in cell cycle is crucial for the deletion event. In the case of administering a high dose of antigen, excessive antigen triggers apoptotic cell death of the autoreactive lymphocytes [13].

Ignorance. Immune ignorance to self antigen appears to occur in those situations in which autoreactive T cells are present and are not tolerant, but simply ignore the self antigen. This ignorance may be due to the fact that the autoreactive T cell expresses a TCR with low affinity for the MHC-peptide complex and has escaped central and peripheral deletion, or ignorance may be due to the inaccessibility of the self antigen.

There are two examples in transgenic models in which ignorance appears to play a role in the lack of CD8+ T cell reactivity. Ohashi et al. [18] and Oldstone et al. [19] have demonstrated that mice transgenic for both a viral glycoprotein or nucleoprotein expressed in pancreatic β-cells and a specific TCR α/β heterodimer, failed to induce tolerance in the transgenic T cells or cause insulitis or diabetes. However, the ignorant state of the CD8+ cell (the lack of autoreactivity) can be broken and the β-cells can be destroyed by infection with live virus. The second example involves a transgenic mouse line which expresses H-2K^b in the β-cells, a TCR specific for H-2K^b in the T cells and IL-2 in the β-cells, introduced by mating between individual transgenic lines [20]. The double transgenic mice which express H-2K^b and the TCR specific for H-2K^b are not tolerant to H-2K^b and fail to develop insulitis. However, the ignorance was reversed by local stimulation by IL-2 (produced by the IL-2 transgene in the β-cells) and the activation of an anti-H-2K^b response resulted in autoimmune damage leading to overt diabetes. These data suggest that immune ignorance can be broken by changes within the microenvironment or in the form of the antigen.

The second possibility for ignorance results from anatomical sequestration of potential autoantigens such as lens or sperm antigens. In addition, some autoantigens may not be expressed on the surface of cells, or are not processed and/or presented via a particular MHC processing/presenting pathway. The limited accessibility or inaccessibility of certain self antigens may lead to immune ignorance rather than tolerance.

Immune Regulation/Suppression
As mentioned above, due to an incomplete deletion process in the thymus, mature autoreactive T cells are detected in the periphery yet remain in an inactive state. In addition to unresponsiveness, ignorance or deletion, the mechanism known as immune regulation or suppression has long been considered to have an important role in maintaining self tolerance in the periphery. Recently, it has become more evident that immune suppression may be mediated by subsets of T cells which produce specific cytokines which in turn down-regulate or inactive the potential autoreactive lymphocytes. For example, cytokines produced by monocytes or lymphocytes can determine and/or switch the type of T helper response towards either a Th1 response with IL-12 and IFN-γ

production [21–23] or towards a Th2 response with IL-4 and IL-10 production [24, 25]. The type of T helper response may, as a result, be a critical determinant in an autoimmune process. A predominant Th1 response may accelerate insulin-dependent diabetes mellitus (IDDM) or experimental autoimmune encephalomyelitis (EAE) since these diseases appear to be mediated by Th1 cells [26]. In contrast, a Th2 response may suppress a Th1 response, and in other diseases may help autoreactive B cells to produce autoantibody and enhance antibody-mediated autoimmune diseases such as autoimmune lupus.

Transforming growth factor-β (TGF-β) is a strongly inhibitory cytokine for T cell and other inflammatory responses. Stimulation of TGF-β production can result in local suppression of T-cell activity [27]. This has been shown in cases of induction of oral tolerance in EAE in rodents and in TGF-β mutant mice [28, 29].

Despite the fact that such control and regulatory mechanisms exist to ensure self tolerance, autoimmune diseases continue to occur, suggesting that multiple factors may be involved in the triggering of an autoimmune response. This has led a search for various different therapeutic strategies to prevent and/or treat autoimmune disease in animal models.

Selective Targets for Immunointervention

Cell Adhesion Molecules

It is still unclear how the interaction between genetic and environmental factors triggers an autoimmune response leading to specific destruction of the target organ. It is clear, however, that recruitment and local accumulation of self-reactive lymphocytes is essential for a destructive inflammatory process. This common lymphocyte homing pathway to the target organ provides a unique opportunity for therapeutic intervention.

The homing of lymphocytes or other inflammatory cells such as neutrophils to a given tissue is a highly regulated process involving different homing receptors on lymphocytes and their corresponding ligands known as vascular or mucosal addressins on high endothelial venules or on cellular matrix [30, 31]. Functionally different lymphocytes such as memory versus naive T cells express different sets of homing receptors [32, 33]. The selective interaction of different homing receptors and/or addressins may direct lymphocyte homing to different tissues.

Based upon a multistep model of lymphocyte adhesion [34], in which selectins (homing receptors) trigger the initial attachment or 'rolling', and integrin receptors mediate stronger adhesion and eventual migration of lymphocytes through the venule endothelium (fig. 1), we designed experiments to

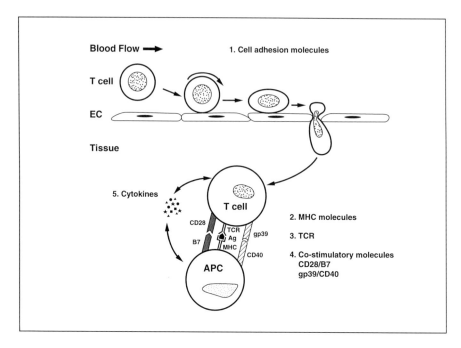

Fig. 1. Selective targets for immunotherapy in autoimmune disease. Autoimmune destruction of target tissues involves a cascade of cellular and molecular interactions which provides a basis for designing potential immunointervention procedures.

address whether lymphocytic invasion of the pancreatic islets can be selectively blocked by interruption of the lymphocyte adhesion process. In initial experiments, using an adoptive transfer model of autoimmune diabetes in which acute disease can be induced by transfer of splenic T cells from diabetic animals to young nondiabetic mice, three homing receptors were chosen as targets namely *L*-selectin, and β_2- and α_4-integrins. The experiments were carried out by treating diabetic spleen cell transfer recipients with monoclonal antibodies (mAbs) specific for these selected homing receptors. It was found that anti-α_4 or anti-*L*-selectin treatment resulted in prevention of transfer of diabetes, while anti-β_2-integrin failed to show a significant effect [35]. Upon further analysis, the protection mediated by anti-*L*-selectin or anti-α_4-integrin mAbs was shown possibly to be due to blockade of lymphocytic infiltration of the islets rather than to deletion of autoreactive lymphocytes or inhibition of the anti-β-islet cell autoimmune response [36]. More recently, we have been able to show that short-term (1 month) treatment of neonatal mice with anti-α_4-integrin mAb can induce long-term (more than 7 months) prevention of *spontaneous* development of IDDM in normal NOD mice. Interestingly, the inhibition of lymphocyte homing by blocking L-selection or α_4-integrin homing receptors seems to be tissue specific since lymphocytic infiltration of the salivary glands (sialitis) was not affected by anti-L-selectin or anti-α_4-integrin treatment. Furthermore, the response to foreign antigens such as chicken ovalbumin in the anti-α_4-integrin-treated mice was not affected [Yang et al.: In prep.]. These data demonstrate that prevention of lymphocyte homing to the target organ is a feasible (if cumbersome) approach for prevention of spontaneous development of autoimmune disease such as IDDM. A similar approach has been utilized to treat other *induced* organ-specific autoimmune diseases such as collagen type II arthritis and EAE in rodents [37, 38].

Cytokines and Cytokine Receptors

Despite the fact that cytokines are functionally redundant and the majority of cytokines have pleiotropic effects on a variety of tissues, it has become increasingly evident that control and neutralization of cytokine activity may be a promising approach for treatment of autoimmune disease.

Tumor Necrosis Factor-α (TNF). TNF is one of the most abundant cytokines found in an inflammatory lesion, suggesting that this cytokine may be a major mediator of autoimmune destruction. It has been shown that neutralization of TNF activitiy by mAb treatment leads to the prevention of collagen-induced arthritis and EAE in mice [39, 40]. Recently, Feldman et al. [41] have demonstrated that treatment with an anti-TNF antibody resulted in alleviation of rheumatoid arthritis in patients [41]. However, the role of TNF in the pathogenesis of type I diabetes remains controversial. On the one hand, an early study has shown that TNF is produced by islet infiltrating CD4+ T cells and that TNF in the presence of other cytokines such as IL-1 or IFN-γ, is cytotoxic to pancreatic islets in vitro [42, 43]. These data support the notion that TNF is involved in the destruction of the β-islet cells. On the other hand, however, it has been found that macrophages from autoimmune diabetes-prone NOD mice or BB rats have a reduced capability to produce TNF in vitro upon stimulation with LPS [44, 45]. It has been hypothesized that the decrease of TNF production in these animals may in some way be associated with disease susceptibility. A further study has demonstrated that treatment of adult NOD mice with TNF protects animals from the development of spontaneous IDDM, supporting this hypothesis [44, 46].

Moreover, it has recently been found that TNF is constitutively produced in the thymus prior to birth and 2–3 weeks after birth [47] and that blocking TNF activity by a polyclonal anti-TNF antiserum leads to thymic and splenic atrophy and lymphopenia [48]. These findings clearly indicate that TNF plays a crucial role in the development of lymphoid tissues.

These interesting and conflicting findings regarding the role of TNF in IDDM prompted us to carry out a series of experiments to gain further insight into the function of TNF in the development of autoimmunity, and to elucidate the mechanisms by which TNF affects the diabetogenic process. We treated neonatal female NOD mice with TNF or a mAb specific for TNF for 3 weeks. Surprisingly, mice treated with TNF developed an earlier onset of disease and 100% incidence compared to 80% in control mice. In contrast, none of the mice treated with anti-TNF mAb developed IDDM even after 30 weeks of age. Autoimmune responses to a panel of β-islet cell autoantigens were enhanced by treatment with TNF but significantly downregulated by anti-TNF treatment suggesting that TNF may function by regulating anti-β-cell autoimmunity. Interestingly, the effect of TNF on IDDM is strikingly age dependent. Treatment with TNF initiated prior to 3 weeks of age led to acceleration of the diabetogenic process while TNF treatment given after 3 weeks of age *delayed* the onset of disease. These data indicate that TNF may induce diverse effects on IDDM depending upon the developmental stage of the immune system. Taken together, we are convinced that TNF does have an important role in the development of autoimmunity as well as in the immunopathogenesis of IDDM and that the protective effect of anti-TNF treatment may have implications for development of a cytokine or cytokine receptor based prevention of human IDDM [49].

Recently, the possibility of developing cytokine- or cytokine-receptor-based immunotherapies in autoimmune and inflammatory diseases has been extensively explored. For example, treatment with recombinant IL-2 or IL-4 protects NOD mice from diabetes, which may result from a reversal and restoration of decreased T-cell proliferative responses, or possibly due to switching of Th1 to Th2 responses by IL-4 treatment [50, 51]. Moreover, for systemic autoimmunity, it has been shown that neutralization of IL-10 activity protects autoimmune lupus in (NZB/W)F1 mice, which appeared to be due to upregulation of endogenous TNF [52].

T-Cell Receptors
Acha-Orbea et al. [53] have demonstrated that the encephalitogenic T-cell clones from PL/J mice (I-A^u) utilize limited TCR Vβ gene segments (predominantly Vβ8.2) and that depletion of the T cells bearing the TCR Vβ8.2 in PL/J mice results in amelioration of autoimmune EAE [53]. However, questions still remain as to whether pathogenic autoreactive T cells in various autoimmune disorders with different target organs or tissues use restricted TCR Vβ segments, and whether this kind of limited TCR usage exists in any human autoimmune disorder [54].

Antigen-Specific Intervention

Depending on the target organ or tissue involved, autoimmune diseases can be classified as systemic or organ specific. For organ-specific diseases, identification of the tissue-specific autoantigen(s) is crucial for the development of an antigen-specific therapy. However, for most human autoimmune diseases, the target autoantigen(s) remains unidentified. Recently, several autoantigens have been identified as targets of the autoimmune process in IDDM in man and in the NOD mouse. This has permitted analysis of the immune response to a panel of putative β-islet cell autoantigens and, in turn, has led to the use of these autoantigens in an attempt to re-establish self tolerance to the β-cells in NOD mice.

To better understand the initial triggering events induced by autoantigens, cDNAs encoding a panel of murine β-cell autoantigens including the two isoforms of glutamic acid decarboxylase (GAD) of 65 and 67 kD, carboxypeptidase-H (CPH), peripherin and the 60-kD heat-shock protein (hsp60) were cloned and the recombinant protein antigens were expressed in baculovirus or *Escherichia coli* expression systems. Both T- and B-cell responses to these autoantigens were detected in diabetes-prone NOD mice, but not in nonautoimmune prone mice such as BALB/c or B10.GD. Interestingly, the responses to the different autoantigens appear temporal. The first detectable response to GAD65 was detected in animals at 4 weeks of age, an age which correlates with the onset of insulitis [55]. Responses to the rest of the antigens developed later, between 6 and 8 weeks of age. These data raised the question whether GAD65 has a critical role in the early development of anti-β-cell autoimmunity.

To address this point, GAD65 antigen was injected intrathymically into 3-week-old NOD mice in order to induce antigen-specific tolerance to GAD65. A single injection of GAD65 resulted in a marked reduction of T-cell proliferation to GAD65 and to the remainder of the panel of β-cell autoantigens and also resulted in prevention of the spontaneous development of IDDM. Similar intrathymic injection of peripherin, however, failed to prevent IDDM. The fact that intrathymic injection of GAD65 led to induction of 'tolerance' not only to GAD65 but also to the other autoantigens, suggests that the protective effect may not simply be due to the induction of an antigen-specific tolerance but may also involve immune regulation/suppression [55]. This is borne out by the observation that antibody responses specific for GAD65 and the other antigens persist in the protected animals. These data and those of Kaufman et al. [56] suggest that GAD65 may have an important role early in the development of IDDM and that induction of an altered response, or tolerance to autoantigens is a feasible approach to prevent autoimmune diabetes.

In order to establish tolerance and prevent autoimmunity, studies utilizing various experimental models have revealed that there are several ways to deliver antigens such as intravenous, oral, intrathymic or neonatal intraperitoneal administration. However, it should be stressed that the antigen and the route utilized to introduce antigen may result in differing ability to prevent the onset of disease and may even sometimes exacerbate the condition. Thus, there are important issues regarding the dose, the form of antigen, the route utilized and the timing of administration to induce or re-establish tolerance in humans. Comprehensive studies are required to address these issues before we can establish a safe, efficient and antigen-based specific immunotherapy for autoimmune disease [57].

Costimulatory Molecules

In addition to antigen receptor-mediated signaling, a costimulatory signal is required for T-cell activation and for B-cell maturation. The lack of a costimulatory signal may result in T cells entering an anergic state and B cells being unable to undergo immunoglobulin class switching [9, 58]. Based on these observations, it may be possible to interrupt an autoimmune response by blocking the interaction between a costimulatory molecule and its ligand. Two such examples exist. In NOD mice, treatment of young animals at 2 weeks of age with either a mAb specific for CD28, a costimulatory molecule on T cells, or with a soluble CTLA-4Ig chimeric molecule which binds to B7, the ligand for CD28 or CTLA-4 on APC, resulted in prevention of IDDM [Bluestone et al., unpubl. data]. In an antibody-mediated autoimmune disease, collagen type II-induced arthritis (CIA) in mice, administration of a mAb specific for gp39, a T-cell ligand of B-cell costimulatory molecule CD40, led to downregulation of autoantibody response and inhibition of CIA [59]. This approach may have the potential to selectively downregulate an ongoing antigen-specific autoimmune response.

Concluding Remarks

Although autoimmunity is manifested in a variety of different clinical diseases involving different target organs, breakdown of self tolerance and an inflammatory response appear to be common pathways leading to the destruction of the target organs. In general, two major approaches are currently being used or examined to prevent and/or treat autoimmunity (fig. 1). The first approach is to reestablish self tolerance or to induce anergy to the target antigen(s) through the introduction of specific autoantigens. The second approach is to selectively interrupt the common pathway(s) required for inflammatory destruction of the target organ, such as blocking leukocyte homing or immunosuppression. Success in developing more efficient therapeutic procedures to treat autoimmune diseases in humans may need combinations of one or more of these approaches.

References

1   Ehrlich P: On immunity with special reference to cell life. Proc R Soc Lond 1900;66:424–448.
2   Sinha AA, Lopez MT, McDevitt HO: Autoimmune diseases: The failure of self tolerance. Science 1990;248:1380–1388.
3   Janeway CAJ: The immune system evolved to discriminate infectious nonself from noninfectious self. Immunol Today 1992;13:11–16.
4   Kappler JW, Roehm N, Marrack P: T cell tolerance by clonal elimination in the thymus. Cell 1987;49:273–280.
5   von Boehmer H: Developmental biology of T cells in T cell-receptor transgenic mice. Ann Rev Immunol 1990;8:531–556.
6   Nossal GJV: Negative selection of lymphocytes. Cell 1994;76:229–239.
7   Hammerling GJ, Schonrich G, Ferber I, Arnold B: Peripheral tolerance as a multi-step mechanism. Immunol Rev 1993;133:93–104.
8   Jenkins MK, Ashwell JD, Schwartz RH: Allogeneic non-T spleen cells restore the responsiveness of normal T cell clones stimulated with antigen and chemically modified antigen-presenting cells. J Immunol 1988;140:3324–3330.
9   Schwartz RH: A cell culture model for T lymphocyte clonal anergy. Science 1990;248:1349–1356.
10 Linsley PS, Brady W, Grosmaire LS, Aruffo A, Damle NK, Ledbetter JA: Binding of the B cell activation antigen B7 to CD28 costimulates T cell proliferation and interleukin-2 mRNA accumulation. J Exp Med 1991;173:721–730.
11 Linsley PS, Brady W, Urnes M, Grosmaire LS, Damle NK, Ledbetter JA: CTLA-4 is a second receptor for the B cell activation antigen B7. J Exp Med 1991;174:561–569.
12 Ledbetter JA, Wallace PM, Johnson J, Gibson MG, Greene JL, Singh C, Tepper MA: Immunosuppression in vivo by a soluble form of the CTLA-4 T cell activation molecule. Science 1992; 257:792–795.
13 Critchfield JM, Racke M, Zuniga-Pflucker JC, Cannella B, Raine C, Goverman J, Lenardo MJ: T cell deletion in high antigen dose therapy of autoimmune encephalomyelitis. Science 1994; 263:1139–1143.
14 Webb S, Morris C, Sprent J: Extrathymic tolerance of mature T cells: Clonal elimination as a consequence of immunity. Cell 1990;63:1249–1256.
15 Burstein HJ, Abbas AK: In vivo role of interleukin 4 in T cell tolerance induced by aqueous protein antigen. J Exp Med 1993;177:457–463.
16 DeWit D, Van Mechelen M, Ryelandt M, Figueiredo AC, Abramowicz D, Goldman M, Bazin H, Urbain J, Leo O: The injection of deaggregated gamma globulins in adult mice induces antigen-specific unresponsiveness of T helper type 1 but not type 2 lymphocytes. J Exp Med 1992;175:9–14.
17 Ferber I, Schonrich G, Schenkel J, Mellor AL, Hammerling G, Arnold B: Level of peripheral T cell tolerance induced by different doses of tolerogen. Science 1994;263:674–676.
18 Ohashi PS, Oehen S, Burki K, Pircher H, Ohashi CT, Odermatt B, Malissen B, Zinkernagel RM, Hengartner H: Cell. Ablation of tolerance and induction of diabetes by virus infection in viral antigen transgenic mice. Cell 1991; 65:305–317.
19 Oldstone MB, Nerenberg M, Southern P, Price J, Lewicki H: Virus infection triggers insulin-de-

pendent diabetes mellitus in a transgenic model: Role of anti-self (virus) immune response. Cell 1991;65:319–331.

20 Miller JFAP, Heath WR: Self-ignorance in the peripheral T cell pool. Immunol Rev 1993;133: 131–150.

21 Mosmann TRR, Coffman RL: Th1 and Th2 cells: Different patterns of lymphokine secretion lead to different functional properties. Ann Rev Immunol 1989;9:145–173.

22 Seder RA, Gazzinelli R, Sher A, Paul WE: IL-12 acts directly on CD4+ T cells to enhance priming for IFN-γ production and diminishes IL-4 inhibition of such priming. Proc Natl Acad Sci USA 1993;90:10188–10192.

23 Trinchieri G: Interleukin-12 and its role in the generation of Th1 cells. Immunol Today 1993; 14:335–337.

24 Swain SL, Weinberg AD, English M, Huston G: IL-4 directs the development of Th2-like helper effectors. J Immunol 1990;145:3796–3806.

25 Seder RA, Paul WE, Davis MM, Fazekas de St. Groth B: The presence of interleukin-4 during in vitro priming determines the cytokine-producing potential of CD4+ T cells from T cell receptor transgenic mice. J Exp Med 1992;176:1091–1098.

26 Liblau RS, Singer SM, McDevitt HO: Th1 and Th2 CD4+ T-cells in the pathogenesis of organ-specific autoimmune diseases. Immunol Today 1994;in press.

27 Sher A, Gazzinelli RT, Oswald IP, Clerici M, Kullberg M, Pearce EJ, Berzofsky JA, Mosmann TR, James SL, Morse HC 3rd: Role of T-cell derived cytokines in the downregulation of immune responses in parasitic and retroviral infection. Immunol Rev 1992;127:183–204.

28 Khoury SJ, Hancock WW, Weiner HL: Oral tolerance to myelin basic protein and natural recovery from experimental autoimmune encephalomyelitis are associated with downregulation of inflammatory of cytokines and differential upregulation of transforming growth factor beta, interleukin 4, and prostaglandin E expression in the brain. J Exp Med 1992;176:1355–1364.

29 Kulkarni AB, Huh CG, Becker D, Geiser A, Lyght M, Flanders KC, Roberts AB, Sporn MB, Ward JM, Karlsson S: Transforming growth factor beta 1 null mutation in mice causes excessive inflammatory response and early death. Proc Natl Acad Sci USA 1993;90:770–774.

30 Springer TA: Adhesion receptors of the immune system. Nature 1990;346:425–434.

31 Springer TA: Traffic signals for lymphocyte recirculation and leukocyte emigration: The multi-step paradigm. Cell 1994;76:301–314.

32 Swain SL, Bradley SL, Croft M, Tankonogy S, Atkins G, Weinberg AD, Duncan DD, Hedrick SM, Dutton RW, Huston G: Helper T-cell subsets: Phenotype, function and the role of lymphokines in regulating their development. Immunol Rev 1991;129:115–144.

33 Picker LJ: Mechanisms of lymphocyte homing. Curr Opin Immunol 1992;4:277–287.

34 Butcher EC: Leukocyte-Endothelial Cell recognition: Three (or more) steps to specificity and diversity. Cell 1991;67:1033–1036.

35 Yang XD, Tisch R, McDevitt HO: Cell adhesion molecules: A selective therapeutic target for alleviation of IDDM. J Autoimmun 1994;in press.

36 Yang XD, Karin N, Tisch R, Steinman L, McDevitt HO: Inhibition of insulitis and prevention of diabetes in NOD mice by blocking L-selectin and VLA-4 adhesion receptors. Proc Natl Acad Sci USA 1993;90:10494–10498.

37 Kakimoto K, Nakamura T, Ishii K, Takashi T, Iigou H, Yagita H, Okumura K, Onoue K: The effect of anti-adhesion molecule antibody on the development of collagen-induced arthritis. Cell Immunol 1992;142:326–337.

38 Yednock TA, Cannon C, Fritz LC, Sanchez-Madrid F, Steinman L, Karin N: Prevention of experimental autoimmune encephalomyelitis by antibodies against α4β1 integrin. Nature 1992; 356:63–66.

39 Williams RO, Feldman M, Maini RN: Anti-tumor necrosis factor ameliorates joint disease in murine collagen-induced arthritis. Proc Natl Acad Sci USA 1992;89:9784–9788.

40 Ruddle NH, Bergman CM, McGrath KM, Lingenheld EG, Grunnet ML, Padula SJ, Clark RB: An antibody to lymphotoxin and tumor necrosis factor prevents transfer of experimental allergic encephalomyelitis. J Exp Med 1990;172:1193–1200.

41 Feldmann M, Brennan FM, Williams RO, Cope AP, Gibbons DL, Katsikis PD, Maini RN:

Evaluation of the role of cytokines in autoimmune disease: the importance of TNF-α in rheumatoid arthritis. Progr Growth Factor Res 1992;4:247–255.

42 Held W, MacDonald HR, Weissman IL, Hess MW, Mueller C: Genes encoding tumor necrosis factor α and granzyme A are expressed during development of autoimmune diabetes. Proc Natl Acad Sci USA 1990;87:2239–2243.

43 Campbell IL, Iscaro A, Harrison LC: IFN-γ and tumor necrosis factor-α. cytotoxicity to murine islets of Langerhans. J Immunol 1988;141:2325–2329.

44 Jacob CO, Aiso S, Michie SA, McDevitt HO, Acha-Orbea H: Prevention of diabetes in nonobese diabetic mice by tumor necrosis factor (TNF): Similarities between TNF-α and interleukin 1. Proc Natl Acad Sci USA 1990;87:968–972.

45 Lapchak PH, Guilbert LJ, Rabinovitch A: Tumor necrosis factor production is deficient in diabetes-prone BB rats and can be corrected by complete Freund's adjuvant: A possible immunoregulatory role of tumor necrosis factor in the prevention of diabetes. Clin Immunol Immunopathol 1992;65:129–134.

46 Satoh J, Seino H, Abo T, Tanaka SI, Shintani S, Ohta S, Tamura K, Sawai T, Nobunaga T, Oteki T, Kumagai K, Toyota T: Recombinant human tumor necrosis factor α suppresses autoimmune diabetes in nonobese diabetic mice. J Clin Invest 1989;84:1345–1348.

47 Giroir BP, Brown T, Beutler B: Constitutive synthesis of tumor necrosis factor in the thymus. Proc Natl Acad Sci USA 1992;89:4864–4868.

48 De Kossodo S, Grau GE, Daneva T, Pointarie P, Fossati L, Ody C, Zapf J, Piguet PF, Gaillard RC, Vassalli P: Tumor necrosis factor α is involved in mouse growth and lymphoid tissue development. J Exp Med 1992;176:1259–1264.

49 Yang XD, Tisch R, Cao ZA, Singer S, Schreiber R, McDevitt HO: The role of tumor necrosis factor-α in the development of autoimmunity in NOD mice. Autoimmunity 1993;15(suppl):42.

50 Zielasek J, Burkart V, Naylor P, Goldstein A, Kiesel U, Kolb H: Interleukin-2-dependent control of disease development in spontaneously diabetic BB rats. Immunology 1990;69:209–214.

51 Rapoport MJ, Jaramillo A, Zipris D, Lazarus A, Serreze DV, Leiter EH, Cyopick P, Danska JS, Delovitch TL: Interleukin-4 reverses T cell proliferative unresponsiveness and prevents the onset of diabetes in nonobese diabetic mice. J Exp Med 1993;178:87–99.

52 Ishida H, Muchamuel T, Sakaguchi S, Andrade S, Menon S, Howard M: Continuous administration of anti-interleukin 10 antibodies delays onset of autoimmunity in NZB/W F1 mice. J Exp Med 1994;179:305–310.

53 Acha-Orbea H, Mitchell DJ, Timmermann L, Wraith DC, Tausch GS, Waldor MK, Zamvil SS, McDevitt HO, Steinman L: Limited heterogeneity of T cell receptors from lymphocytes mediating autoimmune encephalomyelitis allows specific immune intervention. Cell 1988;54:263–273.

54 Wilson DB, Steinman L, Gold DP: The V-region disease hypothesis: New evidence suggests it is probably wrong. Immunol Today 1993;14:376–380.

55 Tisch R, Yang XD, Singer SM, Liblau RS, Fugger L, McDevitt HO: Immune response to glutamic acid decarboxylase correlates with insulitis in non-obese diabetic mice. Nature 1993; 366:72–75.

56 Kaufman DL, Clare-Salzler M, Tian J, Forsthuber T, Ting GSP, Robinson P, Atkinson MA, Sercarz EE, Tobin AJ, Lehmann PV: Spontaneous loss of T cell tolerance to glutamic acid decarboxylase in murine insulin-dependent diabetes. Nature 1993;366:69–72.

57 Tisch R, McDevitt HO: Antigen-specific immunotherapy: Is it a real possibility to combat T-cell mediated autoimmunity. Proc Natl Acad Sci USA 1994;91:437–438.

58 Noelle RJ, Roy M, Shepherd DM, Stamenkovic I, Ledbetter JA, Aruffo A: A 39-kDa protein on activated helper T cells binds CD40 and transduces the signal for cognate activation of B-cells. Proc Natl Acad Sci USA 1992;89:6550–6554.

59 Durie FH, Fara RA, Foy TM, Aruffo A, Ledbetter JA, Noelle RJ: Prevention of collagen-induced arthritis with an antibody to gp39, the ligand for CD40. Science 1993;261:1328–1330.

Xiao-Dong Yang, MD, Department of Microbiology and Immunology,
Stanford University School of Medicine, Stanford, CA 94305 (USA)

Adorini L (ed): Selective Immunosuppression: Basic Concepts and Clinical Applications.
Chem Immunol. Basel, Karger, 1995, vol 60, pp 32–47

....

Immunosuppression in Insulin-Dependent Diabetes mellitus: From Cellular Selectivity towards Autoantigen Specificity

Jean-François Bach, Lucienne Chatenoud

INSERM U 25, Hôpital Necker, Paris, France

Insulin-dependent Diabetes mellitus (IDDM) may be considered as the prototype of T-cell-mediated autoimmune diseases. Evidence has accumulated over the last 10 years to establish, on a firm basis, the autoimmune nature of the disease [1]. Human IDDM is associated with various anti-islet cell autoantibodies and well-defined HLA alleles. The disease may be transferred to nondiabetics by bone marrow transplantation and is slowed down by cyclosporin A. The islets of Langerhans are infiltrated with mononuclear cells including a majority of T cells and macrophages. Even more direct evidence in favor of the autoimmune origin is found in the two models of spontaneous IDDM, the NOD mouse and the BB rat where disease transfer can be demonstrated with T-cell clones and is inhibited by a wide spectrum of immunosuppressive agents. The major pending problem remains the identification of the target autoantigen. Several candidates have been described including glutamic acid decarboxylase (GAD), insulin and heat-shock protein 60 (hsp 60) but no convincing argument has been brought to definitively prove their involvement in the disease pathogenesis.

The objectives of immunointervention in IDDM are fairly straightforward. Preventing the glucose metabolism disturbances that require insulin administration would allow to avoid: the daily constraint of insulin injections and glycemia monitoring, the acute metabolic accident and even more importantly to decrease the rate and severity of the degenerative complications of the disease. A recent study [2] has indeed shown that the strict control of glycemia allowed by intensive insulin therapy decreased the complication risk at the price of increased constraints and frequent hypoglycemia. One may assume that immunoprevention of the disease would reduce this complication risk even more dramatically by avoiding the onset of hyperglycemia episodes. The problem is, however, to achieve such immunoprevention without exposing the patient to the unacceptable risks of drug toxicity and overimmunosuppression.

We shall successively review the main approaches that have been taken in the immunotherapy of IDDM starting from totally nonspecific measures to presumably strictly autoantigen-specific strategies. In this discussion, we shall rely heavily on animal models of the disease inasmuch as only a minority of the agents to be discussed have been tested in man.

Nonspecific Approaches

Immune Modulation
The β-cell lesion evolves in two phases: inflammation and cytolysis leading to atrophy. One possible approach consists of attempting to reduce the inflammatory and cytolytic mechanisms. An increasingly long list of agents has been tested in the animal models including: anti-oxidants; nicotinamide; desferioxamine, and NO inhibitors.

Only nicotinamide has been used so far in human IDDM with still uncertain efficacy in spite of promising early reports. In fact, the dosage shown to be (transiently) efficacious in NOD mice is far higher than that used in man.

Insulin therapy at low dosage has been proposed as a useful therapy to slow down the disease progression and delay the onset of insulin dependency [3–7]. The mechanisms of action of insulin in this setting is still unclear but could include a direct effect on the β-cell (β-cell rest), possibly through decreased autoantigen expression. Other data, however, suggest that insulin could act as the target autoantigen: parenteral immunization of prediabetic NOD mice with insulin or biologically inactive B chain prevents the onset of IDDM [8].

Another nonspecific approach consists of administering various immunoregulatory products such as interleukin-1 [9], poly I-C, anti-interferon-γ monoclonal antibodies [10–12] or immunostimulant preparations, such as streptococcal extracts or complete Freund's adjuvant (CFA) [13, 14]. Most of these agents are only efficacious when given early in the life of NOD mice (before 10 weeks of age) with the possible exception of CFA that has been reported to be also active at later stages. The CFA effect is particularly spectacular. It could involve suppressor cells since it is abrogated by cyclosphosphamide therapy [15]. Its active principle is not known (we have recently shown in our laboratory that it is not the muramyl dipeptide, the adjuvant active component of mycobacteria for vaccines). The mode of action is not clear

Table 1. Immunointervention strategies applied in NOD mice for prevention and/or treatment of spontaneous diabetes

Treatment/drug	Age at beginning of treatment, weeks	Duration, weeks	Follow-up after treatment, weeks	Relapse	References
Treatment of ongoing disease					
Anti-CD3 (145 2C11)	diabetic	5 days	15–20	no (200 days follow-up)	40
ALS	diabetic	2 days	28	no (200 days follow-up)	32
Depleting anti-CD4+depleting anti-CD8	diabetic	2 days	28	no (200 days follow-up)	32
hsp 65 peptide	diabetic	single injection	23	no	61
Prevention of spontaneous diabetes					
Peptide blocking Ag presentation by NOD class II	3	19	27	yes	56
Murine TNFα	9–10	12	34	no	85
Human TNFα	4	19	19	not tested	73
Human IL-1α	9–10	12	34	no	85
OK-432 (streptococcal preparation)	4	19	19	not tested	73
Human lymphotoxin (TNFβ)	4	26	26	not tested	86
Anti-CD3 (145 2C11)	neonates	single injection	32	no	87
Rabbit antimouse Ig	neonates	4	40	no	88
hsp 65/IFA	4	single administration	36	no	59
hsp 65/PBS	4	single administration	36	no	59–60
p277 of hsp 65/IFA	4	single administration	32	no	59–60
CFA	4–10	single administration	36	no	89
Anti-class I Kd (31-3-4S)	4	25	25	not tested	53
Anti-CD4 (GK 1.5)	2 or 8	4	30	no	90
Rapamycin	8	17	33	no at 6 mg/kg 30% at 0.6 mg/kg	31
FK506	5	15	35	no	28
Anti-TCR (H57-597)	8	16	32	no	57
Depleting anti-CD4 (GK 1.5)	12	20	20	not tested	91
Depleting anti-CD8 (53-6.7)	12	20	20	not tested	91
Anti-CD45RA (14.8)	12	20	20	not tested	91
Anti-CD45RA+ depleting anti-CD8 (14.8 + 53-6.7)	12	20	20	not tested	91
Pork insulin s.c.	4	21	22	no in the 1st week follow-up	3
Pork insulin p.o.	5	52	52	not tested	72
Anti-class II (10-3-6)	3	31	31	not tested	54
Cyclosporin A	4–8	23	23	not tested	20
GK1.5	12	25	25	not tested	95

either. It might include, as suggested by recent data, a selective stimulation of TH2 cells. Administration of CFA itself is difficult to envisage in human IDDM but BCG shows a similar protective effect as CFA in NOD mice and is in fact under investigation in human prediabetics [15].

Nonspecific Immunosuppression

Corticosteroids have been used in experimental as well as in human IDDM to prevent the onset of the disease [16]. The effect was not spectacular and its interpretation was rendered difficult by their effects on glucose metabolism. Other nonselective approaches have been used in children, NOD mice and BB rats: total lymphoid irradiation [17] as well as azathioprine [18] and deoxyspergualine [19].

T- Cell Selective Immunosuppression (tables 1, 2)

Cyclosporin A

Cyclosporin A (CsA) is a potent inhibitor of IDDM in NOD mice and BB rats [20–23]. When administered early enough, it totally prevents the onset of the disease. It loses its efficacy, however, when the disease is fully established. Thus, CsA does not prevent the recurrence of diabetes in overtly diabetic NOD mice grafted with syngeneic islet [24]. Additionally, the drug effect is usually transient since the disease most often recurs when the treatment is stopped.

CsA is also efficacious in human IDDM as shown in two placebo-controlled randomized trials [25, 26]. The drug induces long-term remission from insulin dependency at dosages of 7.5 mg/kg/day. Some nephrotoxicity is observed at this dosage but it can be limited by adequate follow-up of creatininemia values and dosage adjustment. In any case, the toxic lesions do not evolve when the treatment is stopped as exemplified by the normal renal function observed 6–8 years posttherapy in patients having shown moderate to severe lesions [27]. Fortunately, there has been no reported case of overimmunosuppression in more than 500 patients thus treated (no opportunistic infection or lymphoma).

CsA therapy is, however, associated with the major problem of remission escape which occurs rapidly (within 1–2 months) when the treatment is stopped but also (after 1–3 years) when the treatment is continued. In the latter case, one must assume that even when the suppression of the autoimmune response is maintained, an autonomous inhibition of β-cell function occurs probably linked to glucotoxicity (some episodes of hyperglycemia are still seen in these patients even when their overall metabolism control is satisfactory as assessed by glycosilated hemoglobin baselines).

Table 2. Immunointervention strategies applied in NOD mice for prevention and/or treatment of transfer- or cyclophosphamide-induced diabetes

Treatment/drug	Age of beginning of treatment weeks	Duration weeks	Follow-up after treatment weeks	References
Transfer-induced diabetes				
Human insulin s.c.	8	4	6	92
Anti CD11b/CD18 (5C6)	8–12	4	4	93
		10 days	4	
Non-depleting anti-CD4 (YTS 177.1)	8–12	2–4	4	35
Anti-IFNγ (R.4.6A2)	8	5	5	11
Murine TNFα	8	2	5	85
Human IL-1α	8	2	5	85
Cyclophosphamide-induced diabetes				
Anti-class I Kd (31-3-4S)	10	4	4	53
Anti-Lyt 2 (3-155)	10	4	4	53
Anti-CD3 (145 2C11)	8	5	6	40
Anti-TCR (H57-597)	8	3	8	57
Anti-IFNγ (R.4.6A2)	8	2	2	12
Anti-IL-6 (6B4)	8	2	2	12
Anti-IFNγ (R.4.6A2)	8	5	5	11
Anti-CD4 (H129.19)	12	2	3	94
Anti-Vβ8 (F23-1)	12	1	2	62

FK506, Rapamycin

These two drugs also bind to immunophillins like CsA. The effect of FK506 has only been assessed in a limited number of experimental studies [28, 29] and in a few patients [30]. It appears similar to that of CsA, although one cannot exclude a greater β-cell toxicity. Rapamycin has only been tested in experimental models and has only been shown to be efficacious in the early phase of the disease [31].

Antilymphocyte Antibodies

Several antilymphocyte antibody preparations have been used in experimental models.

Polyclonal antilymphocyte area (ALS) have been used successfully in the NOD mouse [32] and in the BB rat [33]. In the former case, the effect was still observed in the very late stages of the disease.

Various anti-T-cell monoclonal antibodies have also been used in the animal models. The most thoroughly studied antibodies have been anti-CD4 [34–39] and to a lesser degree anti-CD3 [40], anti-CD28 and anti-CD25 [41] antibodies. The effects of anti-CD3 and anti-CD4 antibodies have appeared particularly interesting and intriguing inasmuch as these antibodies have proven able to induce long-term remission of the disease, thus restoring self tolerance. The mechanisms of such tolerance are still ill defined. It seems to be an active phenomenon since it is broken by cyclophosphamide at doses known to eliminate suppressor T cells but not broken by infusion of diabetogenic T cells.

Only anti-CD3 [42], and anti-CD 25 antibodies have been used in man in a very limited number of patients. The treatment was followed in the case of the anti-CD3 antibody by cyclosporin.

Immunotoxins

An interleukin-2 diphtheria toxin has been successfully used in NOD mice (on diabetes transfer) [43] and in human IDDM [42]. The idea there is to target IL-2 receptor expressing T cells in the islet and kill them selectively by the toxin. Highly promising data obtained in 43 patients have prompted us to perform a randomized trial now under progress. It will be interesting to know in the long term if initial destruction of highly activated T cells, presumably T-cells which are most resistant to the action of CsA, will bring an additional effect to CsA. This strategy should in theory be applicable to prediabetics with the reservation that one would like to verify that such treatment does not also hit suppressor cells that may contribute to protection from the disease.

MHC-Based Therapy

The MHC molecules represent with the T-cell receptor and the antigenic peptide a major component of the molecular complex of antigen recognition. MHC class I molecules present peptides to CD8 T cells while MHC class II molecules present antigens to CD4 T cells. Three approaches have been taken in the NOD mouse to prevent disease onset by targetting MHC molecules: (1) MHC transgenic mice; (2) anti-MHC monoclonal antibodies, and (3) MHC-blocking peptides.

MHC Transgenic Mice

NOD mice are characterized by the presence of a unique I-A molecule and by the absence of I-E molecules. Several groups of investigators have produced I-A and I-E transgenic NOD mice. Introduction of numerous copies of non-NOD I-A molecules has various effects depending on the nature of the

I-A gene utilized. NOD mice are protected by I-A^d [44] and I-A^k [45–47]. They are also protected by the introduction of an I-A^{NOD} molecule mutated at position 56 of the I-Aβ chain [48] but not at position 57 [48] although this latter position has been reported to be a critical position for diabetes susceptibility. Depending on the investigator, I-E expression protects [48–50] or does not protect [51] NOD mice from diabetes. In all cases where protection is obtained, one may wonder whether such protection is due to uneffective presentation of the diabetogenic peptides by MHC molecules or to the induction of suppressor mechanisms that prevent the function of T cells having recognized the peptide in the context of the initial NOD mouse MHC molecules (still expressed in these transgenic mice) or even by the transgenic MHC molecules. Evidence in favor of the latter mechanism has recently been obtaind both in the I-A^d [44] and the I-A^k [45] transgenes. Although these transgenic mice are protected from diabetes, such mice become diabetic after cyclophosphamide administration and their spleen cells can transfer diabetes in sublethally irradiated syngeneic mice. Additionally, in cotransfer experiments, it was shown that such spleen cells could inhibit the transfer normally obtained with spleen cells from non-transgenic NOD mice. Thus, it would appear that MHC molecules can drive the appearance of regulatory cells that inhibit the diabetogenic process.

One might also mention here the recent report by Katz et al. [52] that NOD mice backcrossed to MHC class I deprived mice (after homologous recombination) do not show diabetes.

Anti-MHC Monoclonal Antibodies

Anti-class I [53] and anti-class II [54] monoclonal antibodies prevent the onset of diabetes in NOD mice. Anti-class II antibodies have also been shown to be active in the BB rat [55]. The action of such anti-class II MHC monoclonals has been studied in some detail in the NOD mouse. A long lasting effect is only obtained when the treatment is applied early (before 8 weeks of age). The antibody induces a permanent state of tolerance persisting over 18 months of age [our unpubl. observations]. Interestingly, this tolerance is not broken down by cyclophosphamide treatment at any age, at variance with anti-CD3- or anti-CD4-induced tolerance. IR regulatory mechanisms may be involved since the protection afforded to anti-class II-treated young NOD mice can be transferred to syngeneic nonantibody-treated recipients by CD4 T cells collected from the class II-treated tolerant mice [54]. This mechanism is more likely than the MHC blockade hypothesis previously presented.

Blocking Peptides

The blocking of a number of T-cell immune responses by MHC-blocking peptides has been reported in several models including autoimmune experimental allergic encephalomyelitis. It was tempting to use it in NOD mice, with the difficulty, that the triggering autoantigen is not known with certainty in this model. Prevention of diabetes has been obtained after early administration of peptides previously shown to bind to NOD MHC class II molecules [56]. The effect was only demonstrated, however, as a prevention treatment, not as a curative one (administered once the disease was overtly declared) which potentially limitates its clinical application.

T-Cell Receptor-Based Immunointervention

Anti-TCR and Anti-CD3 Monoclonal Antibodies

Anti-TCR (T-cell receptor) monoclonal antibodies [57] may prevent diabetes onset in NOD mice and even induce transient regression of overtly established disease (with correction of hyperglycemia). A similar effect is obtained with anti-CD3 monoclonals [40], with the major difference that the regression of overt diabetes is long lasting, surviving the cessation of therapy. We have thus recently shown that 5 consecutive daily injections of 5 µg of a hamster antimouse CD3 monoclonal induces permanent remission of diabetes inasmuch as the mice are treated shortly after diabetes onset, presumably before too many β-cells are destroyed [40]. In fact, even in these resistant mice the antibody induces tolerance since a syngeneic islet graft is not destroyed by the autoimmune process as it is in non-antibody-treated mice. The mode of action of the anti-CD3 antibody is currently under investigation. Several data argue in favor of the induction of regulatory T cells (perhaps of the TH2 type): (1) the mononuclear cell infiltrate (insulitis) does not disappear but is redistributed to the islet periphery; (2) diabetogenic spleen cells from overtly diabetic mice do not break down the disease remission; (3) a single injection of high-dose cyclophosphamide induces the appearance of diabetes. It remains to be determined if the monoclonal Fc-dependent T-cell activation contributes to tolerance induction (which is not the case for the preventive effect of the antibody which is obtained with F(ab')2 fragments). It is hypothesized but remains to be proven that the tolerance is driven by the islet cell antigens persisting in the pancreas. If this were the case, the approach would not require knowledge of the triggering autoantigen.

T-Cell Vaccination

Only limited experience has been gained in IDDM with T-cell vaccination. One may mention the attempt by Smerdon et al. [58] to prevent diabetes onset by immunizing with spleen cells derived from diabetic NOD mice. One should also mention the interesting studies by Cohen's group using hsp 60-reactive T-cell clones [59, 60]. hsp 65 induces an accelerated and transient form of diabetes in young NOD mice, which ultimately become refractory to the disease. The effect can be obtained with a peptide derived from the hsp 60 sequence. Transfer studies have indicated that disease prevention is due to the action of T cells reacting in a clonotypic fashion to hsp 60 (and the p277) reactive clones. Interestingly, immunization of overtly diabetic mice with the p277 peptide induces diabetes remission [61].

Vβ Inhibition

It has been shown in EAE that the TCR from encephalitogenic clones used restricted Vβ TCR genes at least in certain conditions of sensitization. Numerous studies have been devoted to the recognition of such a putative restriction in the NOD mouse with possible therapeutic applications. Some approaches have suggested the existence of restriction notably the inhibition of diabetes onset after cyclophosphamide therapy by administering an anti-Vβ8.2 monoclonal [62] or by transfer of diabetogenic spleen cells using an anti-Vβ6 antibody [63]. Some studies using PCR having indicated the TCR of intraislet TT cells in young NOD mice islet cells show restriction of Vβ gene usage but other data have only found an intra-islet oligoclonality of TCR sequences without overall pancreatic restriction [Sarukhan, in preparation]. Lastly, the study of T-cell clones has not provided evidence for any type of restriction even when considering selectively pathogenic clones (shown to transfer the disease) [64].

The inconsistency of these results renders problematic the targeting of Vβ gene products in the therapy of IDDM whatever the strategy used: anti-Vβ monoclonal administration of sensitization against Vβ peptides. The results mentioned above with anti-Vβ8.2 and anti-Vβ6 antibodies must be confirmed. In any case, it is difficult to exclude the possibility of an escape mechanism using another Vβ gene than the dominant one as suggested by the occurrence of diabetes in transgenic NOD mice expressing a fixed Vβ gene not derived from diabetogenic clones [65]. The discussion could be extended to other TCR segments (Vα, J) but there is no evidence that this approach would provide better results than Vβ fragments.

Autoantigen-Specific Therapy

Specific unresponsiveness can be induced in young or adult animals against various antigens by injecting them with the said antigen by various routes. This has notably been achieved with intravenous thyroglobulin in the case of experimental allergic thyroiditis [66], a myelin basic protein in Freund's incomplete adjuvant in the case of EAE. The problem is rendered difficult in diabetes by the elusive nature of the autoantigen. There is good evidence, as in other organ-specific autoimmune diseases, that the autoimmune response is driven by the autoantigen (for example, NOD mice deprived of β-cells by alloxan treatment can not longer sustain the survival of diabetogenic T cells as intermediary hosts) [our unpubl. data], but the number and nature of the autoantigens are still essentially unknown. Several interesting candidates have been described, notably glutamic acid decarboxylase (GAD), hsp60, insulin, a 69-kD protein cross-reactive with bovine serum albumin, but none of them has proven to be the IDDM autoantigen.

One approach is to tolerize using whole islets. This has been achieved both in NOD mice and BB rats by grafting islet intrathymically in young animals [67–69]. This very procedure previously known to induce allogeneic tolerance, also induces self tolerance to islet antigens. Its feasibility in man is, however, questionable inasmuch as human thymus becomes atrophic very early in life.

More specific approaches have recently been attempted with GAD. Intravenous or intrathymic injection of GAD in 3-week-old NOD mice induces tolerance to the protein but also prevents the onset of insulitis and diabetes [70, 71]. Interestingly, tolerance spreads to other β-cell antigens (peripherin, hsp 65). It remains to be known whether GAD is the first triggering autoantigen as suggested by the observation that tolerance to the other autoantigen does not protect from diabetes and does not induce spread tolerance to GAD.

Another approach is to administer autoantigens by the oral route. This has been achieved both for insulin [72] and GAD. Interestingly, this tolerance is not specific of the tolerogen since it can be broken down by non antigen specific means such as anti-TGFβ monoclonal antibody. The current hypothesis is that oral autoantigen induces the appearance of suppression focussed to the autoantigen expressing tissue.

Let us also recall the case of hsp 65 discussed above where the antigen-specific tolerance appears to involve T-cell vaccination.

Miscellaneous Approaches

A number of methods have been proposed to prevent the onset of diabetes in NOD mice or BB rats. None of them has been shown to act at late stages of the disease with the possible exception of Freund's complete adjuvant.

Various forms of immunostimulation prevent the onset of diabetes in NOD mice: poly I poly C, streptococcal extracts [73], several viruses such as lymphochoriomeningitis virus [74, 75], lactodehydrogenase virus [76], murine hepatitis virus [77] and perhaps, more interestingly, complete Freund's adjuvant (CFA) [13, 14] or BCG [78]. Other methods aim at nonspecific β-cell protection from cellular aggressions: antioxidants [79], NO inhibitors [80], nicotinamide [81].

Lastly, one may mention various diets such casein-free [82, 83] or essential fatty acid free diets [84] that can provide an interesting prevention of the disease.

Conclusions

A wide variety of methods has been shown to prevent IDDM onset or to induce its regression once it has appeared. Only few of these methods have been used clinically. The concern has emerged that some of the preventive interventions applied very early in life in NOD mice, could not be totally specific and in any case are not applicable to human diabetics or even prediabetics who are only identified at a much later stage of the natural history.

This risk of nonspecific action is important to consider on a theoretical basis (what is the value of the crucial GAD tolerance experiments?). One may assume that it indicates the plasticity of the autoimmune potential at this early age which could be extrapolated for future intervention in subjects presenting predisposing genes and exposed to triggering environmental factors. The methods described in this review represent a wide gradient of selectivity and specificity. One should not totally oppose antigen-specific and non-antigen-specific approaches since it might turn out, as in transplantation immunology, that induction of antigen-specific tolerance will require transient administration of nonspecific immunosuppressive therapy.

References

1   Bach JF: Insulin-dependent diabetes mellitus as an autoimmune disease. Endocr Rev 1994;in press.
2   The Diabetes Control and Complications Trial Research Group: The effect of intensive treatment of diabetes on the development and progression of long-term complications in insulin-dependent diabetes mellitus. N Engl J Med 1993;329:977–986.

3   Atkinson MA, MacLaren NK, Luchetta R: Insulitis and diabetes in NOD mice reduced by pro-
    phylactic insulin therapy. Diabetes 1990;39:933–937.
4   Thivolet CH, Goillot E, Bedossa P, Durand A, Bonnard M, Orgiazzi J: Insulin prevents adop-
    tive cell transfer of diabetes in the autoimmune non-obese diabetic mouse. Diabetologia 1991;
    34:314–319.
5   Gotfredsen CF, Buschard K, Frandsen EK: Reduction of diabetes incidence of BB Wistar rats by
    early prophylactic insulin treatment of diabetes-prone animals. Diabetologia 1985;28:933–935.
6   Bertrand S, de Paepe M, Vigeant C, Yale JF: Prevention of adoptive transfer in BB rats by pro-
    phylactic insulin treatment. Diabetes 1992;41:1273–1277.
7   Gottlieb PA, Handler ES, Appel MC, Greiner DL, Mordes JP, Rossini AA: Insulin treatment
    prevents diabetes mellitus but not thyroiditis in RT6-depleted diabetes resistant BB/Wor rats.
    Diabetologia 1991;34:296–300.
8   Muir A, Luchetta R, Song HY, Peck A, Krischer J, MacLaren N: Insulin immunization protects
    NOD mice from diabetes (abstract). Autoimmunity 1993;15(suppl):58.
9   Formby B, Jacobs C, Dubuc P, Shao T: Exogenous administration of IL-1-alpha inhibits active
    and adoptive transfer autoimmune diabetes in NOD mice. Autoimmunity 1992;12:21–27.
10 Pujol-Borrell R, Todd I, Doshi M, Bottazzo GF, Sutton R, Gray D, Adolf GR, Feldmann M:
    HLA class II induction in human islet cells by interferon-gamma plus tumour necrosis factor or
    lymphotoxin. Nature 1987;326:304–306.
11 Debray-Sachs M, Carnaud C, Boitard C, Cohen H, Gresser I, Bedossa P, Bach JF: Prevention
    of diabetes in NOD mice treated with antibody to murine IFN-gamma. J Autoimmun 1991;
    4:237–248.
12 Campbell IL, Kay TW, Oxbrow L, Harrison LC: Essential role for interferon-gamma and inter-
    leukin-6 in autoimmune insulin-dependent diabetes in NOD/Wehi mice. J Clin Invest 1991;87:
    739–742.
13 Sadelain MW, Qin HY, Lauzon J, Singh B: Prevention of type I diabetes in NOD mice by adju-
    vant immunotherapy. Diabetes 1990;39:583–589.
14 McInerney MF, Pek SB, Thomas DW: Prevention of insulitis and diabetes onset by treatment
    with complete Freund's adjuvant in NOD mice. Diabetes 1991;40:715–725.
15 Shehadeh N, Calcinaro F, Bradley BJ, Bruchlim I, Vardi P, Lafferty KJ: Effect of adjuvant ther-
    apy on development of diabetes in mouse and man. Lancet 1994;343:706–707.
16 Secchi A, Pastore MR, Sergi A, Pontirolli AE, Pozza G: Prednisone administration in recent
    onset type I diabetes. J Autoimmun 1990;3:593–600.
17 Rossini AA, Slavin S, Woda BA, Geisberg M, Like AA, Mordes JP: Total lymphoid irradiation
    prevents diabetes mellitus in the Bio-Breeding/Worcester (BB/W) rat. Diabetes 1984;33:
    543–547.
18 Cook JJ, Hudson I, Harrison LC, Dean B, Colman PG, Werther GA, Warne GL, Court JM:
    Double-blind controlled trial of azathioprine in children with newly diagnosed type I diabetes.
    Diabetes 1989;38:779–783.
19 Nicoletti F, Borghi MO, Meroni PL, Barcellini W, Fain C, DiMarco R, Menta R; Schorlem-
    mer HU, Bruno G, Magro G, Grasso S: Prevention of cyclophosphamide-induced diabetes in
    the NOD/WEHI mouse with deoxyspergualin. Clin Exp Immunol 1993;91:232–236.
20 Mori Y, Suko M, Okudaira H, Matsuba I, Tsuruoka A, Sasaki A, Yokoyama H, Tanase T,
    Shida T, Nishimura M, Terada E, Ikeda Y: Preventive effects of cyclosporin on diabetes in
    NOD mice. Diabetologia 1986;29:244–247.
21 Laupacis A, Stiller CR, Gardell C, Keown P, Dupre J, Wallace AC, Thibert P: Cyclosporin pre-
    vents diabetes in BB Wistar rats. Lancet 1983;1:10–12.
22 Jaworski MA, Honore L, Jewell LD, Mehta JG, McGuire-Clark P, Schouls JJ, Yap WY: Cyclo-
    sporin prophylaxis induces long-term prevention of diabetes, and inhibits lymphocytic infiltra-
    tion in multiple target tissues in the high-risk BB rat. Diabetes Res 1986;3:1–6.
23 Like AA, Dirodi V, Thomas S, Guberski DL, Rossini AA: Prevention of diabetes mellitus in
    the BB/W rat with cyclosporin-A. Am J Pathol 1984;117:92–97.
24 Wang Y, McDuffie M, Nomikos IN, Hao L, Lafferty KJ: Effect of cyclosporine on immunologi-
    cally mediated diabetes in nonobese diabetic mice. Transplantation 1988;46(2suppl):101S–106S.

25   Feutren G, Papoz L, Assan R, Vialettes B, Karsenty G, Vexiau P, Du Rostu H, Rodier M, Sir-
      mai J, Lallemand A, Bach JF: Cyclosporin increases the rate and length of remissions in insu-
      lin-dependent diabetes of recent onset. Results of a multicentre double-blind trial. Lancet 1986;
      ii:119–124.

26   The Canadian-European Randomized Control Trial Group: Cyclosporin-induced remission of
      IDDM after early intervention. Association of 1 yr of cyclosporin treatment with enhanced in-
      sulin secretion. Diabetes 1988;37:1574–1582.

27   Assan R, Timsit J, Feutren G, Bougneres P, Czernichow P, Hannedouche T, Boitard C, Noel
      LH, Mihatsch MJ, Bach JF: The kidney in cyclosporin A-treated diabetic patients: A long-term
      clinicopathological study. Clin Nephrol 1994;41:41–49.

28   Miyagawa J, Yamamoto K, Hanafusa T, Itoh N, Nakagawa C, Otsuka A, Katsura H, Yama-
      gata K, Miyazaki A, Kono N, Tarui S: Preventive effect of a new immunosuppressant FK-506
      on insulitis and diabetes in nonobese diabetic mice. Diabetologia 1990;33(8):503–505.

29   Strasser S, Cearns-Spielman J, Carroll P, Alejandro R: Effect of FK 506 on cyclophosphamide-
      induced diabetes in NOD mice. Diabetes Nutr Metab 1992;5:61–63.

30   Carroll PB, Tzakis AG, Ricordi C, Rilo HR, Abu-El-Magd K, Murase N, Zeng YJ, Alejan-
      dro R, Mintz D, Starzl TE: The use of FK 506 in new-onset type 1 diabetes in man. Transplant
      Proc 1991;23:3351–3353.

31   Baeder WL, Sredy J, Sehgal SN, Chang JY, Adams LM: Rapamycin prevents the onset of insu-
      lin-dependent diabetes mellitus (IDDM) in NOD mice. Clin Exp Immunol 1992;89:174–178.

32   Maki T, Ichikawa T, Blanco R, Porter J: Long-term abrogation of autoimmune diabetes in non-
      obese diabetic mice by immunotherapy with anti-lymphocyte serum. Proc Natl Acad Sci USA
      1992;89:3434–3438.

33   Like AA, Rossini AA, Guberski DL, Appel MC, Williams RM: Spontaneous diabetes mellitus:
      Reversal and prevention in the BB/W rat with antiserum to rat lymphocytes. Science 1979;
      206:1421–1423.

34   Wang Y, Pontesilli O, Gill RG, La Rosa FG, Lafferty KJ: The role of CD4+ and CD8+ T cells in
      the destruction of islet grafts by spontaneously diabetic mice. Proc Natl Acad Sci USA 1991;
      88:527–531.

35   Hutchings P, O'Reilly L, Parish NM, Waldmann H, Cooke A: The use of a non-depleting anti-
      CD4 monoclonal antibody to reestablish tolerance to beta cells in NOD mice. Eur J Immunol
      1992;22:1913–1918.

36   Hayward AR, Schriber M, Cooke A, Waldmann H: Prevention of diabetes but not insulitis in
      NOD mice injected with antibody to CD4. J Autoimmun 1993;6:301–310.

37   Wang Y, Hao L, Gill RG, Lafferty KJ: Autoimmune diabetes in NOD mouse is L3T3 T-lym-
      phocyte dependent. Diabetes 1987;36:535–538.

38   Koike T, Itoh Y, Ishii T, Ito I, Takabayashi K, Maruyama N, Tomioka H, Yoshida S: Preventive
      effect of monoclonal anti-L3T4 antibody on development of diabetes in NOD mice. Diabetes
      1987;36:539–541.

39   Shizuru JA, Taylor-Edwards C, Banks BA, Gregory AK, Fathman CG: Immunotherapy of the
      nonobese diabetic mouse: Treatment with an antibody to T-helper lymphocytes. Science 1988;
      240:659–662.

40   Chatenoud L, Thervet E, Primo J, Bach JF: Anti-CD3 antibody induces long-term remission of
      overt autoimmunity in nonobese diabetic mice. Proc Natl Acad Sci USA 1994:91:123–127.

41   Kelley VE, Gaulton GN, Hattori M, Ikegami H, Eisenbarth G, Strom TB: Anti-interleukin 2
      receptor antibody suppresses murine diabetic insulitis and lupus nephritis. J Immunol 1988;
      140:59–61.

42   Bach JF: Strategies in immunotherapy of insulin-dependent diabetes mellitus. Ann NY
      Acad Sci 1994;in press.

43   Pacheco-Silva A, Bastos MG, Muggia RA, Pankewycz O, Nichols J, Murphy JR, Strom TB, Ru-
      bin-Kelley VE: Interleukin 2 receptor targeted fusion toxin (DAB486-IL-2) treatment blocks
      diabetogenic autoimmunity in non-obese diabetic mice. Eur J Immunol 1992;22:697–702.

44   Singer SM, Tisch R, Yang XD, McDevitt HO: An Abd transgene prevents diabetes in nonobese
      diabetic mice by inducing regulatory T cells. Proc Natl Acad Sci USA 1993;90:9566–9570.

45   Slattery RM, Kjer-Nielsen L, Allison J, Charlton B, Mandel TE, Miller JF: Prevention of diabetes in non-obese diabetic I-Ak transgenic mice. Nature 1990;345:724–726.

46   Miyazaki T, Uno M, Uehira M, Kikutani H, Kishimoto T, Kimoto M, Nishimoto H, Miyazaki J, Yamamura K: Direct evidence for the contribution of the unique I-ANOD to the development of insulitis in non-obese diabetic mice. Nature 1990;345:722–724.

47   Slattery RM, Miller JF, Heath WR, Charlton B: Failure of a protective major histocompatibility complex class II molecule to delete autoreactive T cells in autoimmune diabetes. Proc Natl Acad Sci USA 1993;90:10808–10810.

48   Lund T, O'Reilly L, Hutchings P, Kanagawa O, Simpson E, Gravely R, Chandler P, Dyson J, Picard JK, Edwards A, Kioussis D, Cooke A: Prevention of insulin-dependent diabetes mellitus in non-obese diabetic mice by transgenes encoding modified I-A beta-chain or normal I-E alpha-chain. Nature 1990;345:727–729.

49   Nishimoto H, Kikutani H, Yamamura K, Kishimoto T: Prevention of autoimmune insulitis by expression of I-E molecules in NOD mice. Nature 1987;328:432–434.

50   Bohme J, Schuhbaur B, Kanagawa O, Benoist C, Mathis D: MHC-linked protection from diabetes dissociated from clonal deletion of T cells. Science 1990;249:293–295.

51   Podolin PL, Pressey A, Delarato NH, Fischer PA, Peterson LB, Wicker LS: I-E+ nonobese diabetic mice develop insulitis and diabetes. J Exp Med 1993;178:793–803.

52   Katz JD, Wang B, Haskins K, Benoist C, Mathis D: Following a diabetogenic T cell from genesis through pathogenesis. Cell 1993;74:1089–1100.

53   Taki T, Nagata M, Ogawa W, Hatamori N, Hayakawa M, Hari J, Shii K, Baba S, Yokono K: Prevention of cyclophosphamide-induced and spontaneous diabetes in NOD/Shi/Kbe mice by anti-MHC class I Kd monoclonal antibody. Diabetes 1991;40:1203–1209.

54   Boitard C, Bendelac A, Richard MF, Carnaud C, Bach JF: Prevention of diabetes in nonobese diabetic mice by anti-I-A monoclonal antibodies: Transfer of protection by splenic T cells. Proc Natl Acad Sci USA 1988;85:9719–9723.

55   Boitard C, Michie S, Serrurier P, Butcher GW, Larkins AP, McDevitt HO: In vivo prevention of thyroid and pancreatic autoimmunity in the BB rat by antibody to class II major histocompatibility complex gene products. Proc Natl Acad Sci USA 1985;82:6627–6631.

56   Hurtenbach U, Lier E, Adorini L, Nagy ZA: Prevention of autoimmune diabetes in non-obese diabetic mice by treatment with a class II major histocompatibility complex-blocking peptide. J Exp Med 1993;177:1499–1504.

57   Sempe P, Bedossa P, Richard MF, Villa MC, Bach JF, Boitard C: Anti-alpha/beta T cell receptor monoclonal antibody provides an efficient therapy for autoimmune diabetes in nonobese diabetic (NOD) mice. Eur J Immunol 1991;21:1163–1169.

58   Smerdon RA, Peakman M, Hussain MJ, Vergani D: Lymphocyte vaccination prevents spontaneous diabetes in the non-obese diabetic mouse. Immunology 1993;80:498–501.

59   Elias D, Markovits D, Reshef T, van der Zee R, Cohen IR: Induction and therapy of autoimmune diabetes in the non-obese diabetic (NOD/Lt) mouse by a 65-kDa heat shock protein. Proc Natl Acad Sci USA 1990;87:1576–1580.

60   Elias D, Reshef T, Birk OS, van der Zee R, Walker MD, Cohen IR: Vaccination against autoimmune mouse diabetes with a T-cell epitope of the human 65-kDa heat shock protein. Proc Natl Acad Sci USA 1991;88:3088–3091.

61   Elias D, Cohen IR: Peptide therapy for diabetes in NOD mice. Lancet 1994;343:704–706.

62   Bacelj A, Charlton B, Mandel TE: Prevention of cyclophosphamide-induced diabetes by anti-V beta 8 T-lymphocyte-receptor monoclonal antibody therapy in NOD/Wehi mice. Diabetes 1989;38:1492–1495.

63   Edouard P, Thivolet C, Bedossa P, Olivi M, Legrand B, Bendelac A, Bach JF, Carnaud C: Evidence for a preferential V beta usage by the T cells which adoptively transfer diabetes in NOD mice. Eur J Immunol 1993;23:727–733.

64   Candeias S, Katz J, Benoist C, Mathis D, Haskins K: Islet-specific T-cell clones from non-obese diabetic mice express heterogeneous T-cell receptors. Proc Natl Acad Sci USA 1991;88:6167–6170.

65 Lipes MA, Rosenzweig A, Tan KN, Tanigawa G, Ladd D, Seidman JG, Eisenbarth GS: Progression to diabetes in nonobese diabetic (NOD) mice with transgenic T cell receptors. Science 1993;259:1165–1169.

66 Parish NM, Rayner D, Cooke A, Roitt IM: An investigation of the nature of induced suppression to experimental autoimmune thyroiditis. Immunology 1988;63:199–203.

67 Gerling IC, Serreze DV, Christianson SW, Leiter EH: Intrathymic islet cell transplantation reduces beta-cell autoimmunity and prevents diabetes in NOD/Lt mice. Diabetes 1992;41: 1672–1676.

68 Koevary SB, Blomberg M: Prevention of diabetes in BB/Wor rats by intrathymic islet injection. J Clin Invest 1992;89:512–516.

69 Posselt AM, Barker CF, Friedman AL, Naji A: Prevention of autoimmune diabetes in the BB rat by intrathymic islet transplantation at birth. Science 1992;256:1321–1324.

70 Kaufman DL, Clare-Salzler M, Tian J, Forsthuber T, Ting GSP, Robinson P, Atkinson MA, Sercarz EE, Tobin AJ, Lehmann PV: Spontaneous loss of T-cell tolerance to glutamic acid decarboxylase in murine insulin-dependent diabetes. Nature 1993;366:69–72.

71 Tisch R, Yang XD, Singer SM, Liblau RS, Fugger L, McDevitt HO: Immune response to glutamic acid decarboxylase correlates with insulitis in non-obese diabetic mice. Nature 1993;366: 72–75.

72 Zhang ZJ, Davidson L, Eisenbarth G, Weiner HL: Suppression of diabetes in nonobese diabetic mice by oral administration of porcine insulin. Proc Natl Acad Sci USA 1991;88:10252–10256.

73 Satoh J, Seino H, Abo T, Tanaka S, Shintani S, Ohta S, Tamura K, Sawai T, Nobunaga T, Oteki T, Kumagai K, Toyota T: Recombinant human tumor necrosis factor alpha suppress autoimmune diabetes in nonobese diabetic mice. J Clin Invest 1989;84:1345–1348.

74 Oldstone MB: Viruses as therapeutic agents. I. Treatment of nonobese insulin-dependent diabetes mice with virus prevents insulin-dependent diabetes mellitus while maintaining general immune competence. J Exp Med 1990;171:2077–2089.

75 Oldstone MB, Ahmed R, Salvato M: Viruses as therapeutic agents. II. Viral reassortants map prevention of insulindependent diabetes mellitus to the small RNA of lymphocytic choriomeningitis virus. J Exp Med 1990;171:2091–2100.

76 Takei I, Asaba Y, Kasatani T, Maruyama T, Watanabe K, Yanagawa T, Saruta T, Ishii T: Suppression of development of diabetes in NOD mice by lactate dehydrogenase virus infection. J Autoimmun 1992:665–673.

77 Wilberz S, Partke HJ, Dagnaes-Hansen F, Herberg L: Persistent MHV (mouse hepatitis virus) infection reduces the incidence of diabetes mellitus in non-obese diabetic mice. Diabetologia 1991;34:2–5.

78 Yagi H, Matsumoto M, Kishimoto Y, Makino S, Harada M: Possible mechanism of the preventive effect of BCG against diabetes mellitus in NOD mouse. II. Suppression of pathogenesis by macrophage transfer from BCG-vaccinated mice. Cell Immunol 1991;138:142–149.

79 Rabinovitch A, Suarez WL, Power RF: Combination therapy with an antioxidant and a corticosteroid prevents autoimmune diabetes in NOD mice. Life Sci 1943;51:1937–1943.

80 Corbett JA, Mikhael A, Shimizu J, Frederick K, Misko TP, McDaniel ML, Kanagawo O, Unanue ER: Nitric oxide production in islets from nonobese diabetic mice: Aminoguanidine-sensitive and -resistant stages in the immunological diabetic process. Proc Natl Acad Sci USA 1993;90:8992–8995.

81 Yamada K, Nonaka K, Hanafusa T, Miyazaki A, Toyoshima H, Tarui S: Preventive and therapeutic effects of large-dose nicotinamide injections on diabetes associated with insulitis. An observation in nonobese diabetic (NOD) mice. Diabetes 1982;31:749–753.

82 Elliott RB, Reddy SN, Bibby NJ, Kida K: Dietary prevention of diabetes in the non-obese diabetic mouse. Diabetologia 1988;31:62–64.

83 Issa-Chergui B, Guttmann RD, Seemayer TA, Kelley VE, Colle E: The effect of diet on the spontaneous insulin dependent diabetic syndrome in the rat. Diabetes Res 1988;9:81–86.

84 Lefkowith J, Schreiner G, Cormier J, Handler ES, Driscoll HK, Greiner D, Mordes JP, Rossini AA: Prevention of diabetes in the BB rat by essential fatty acid deficiency. Relationship between physiological and biochemical changes. J Exp Med 1990;171:729–743.

85 Jacob CO, Aiso S, Michie SA, McDevitt HO, Acha-Orbea H: Prevention of diabetes in non-obese diabetic mice by tumor necrosis factor (TNF): Similarities between TNF-alpha and interleukin 1. Proc Natl Acad Sci USA 1990;87:968–972.

86 Seino H, Takahashi K, Satoh J, Zhu XP, Sagara M, Masuda T, Nobunaga T, Funahashi I, Kajikawa T, Toyota T: Prevention of autoimmune diabetes with lymphotoxin in NOD mice. Diabetes 1993;42:398–404.

87 Hayward AR, Shreiber M: Neonatal injection of CD3 antibody into nonobese diabetic mice reduces the incidence of insulitits and diabetes. J Immunol 1989;143:1555–1559.

88 Forsgren S, Andersson A, Hillorn V, Soderstrom A, Holmberg D: Immunoglobulin-mediated prevention of autoimmune diabetes in the non-obese diabetic (NOD) mouse. Scand J Immunol 1991;34:445–451.

89 Sadelain MW, Qin HY, Lauzon J, Singh B: Prevention of type I diabetes in NOD mice by adjuvant immunotherapy. Diabetes 1990;39:583–589.

90 Kurasawa K, Sakamoto A, Maeda T, Sumida T, Ito I, Tomioka H, Yoshida S, Koike T: Short-term administration of anti-L3T4 MoAb prevents diabetes in NOD mice. Clin Exp Immunol 1993;91:376–380.

91 Sempe P, Ezine S, Marvel J, Bedossa P, Richard MF, Bach JF, Boitard C: Role of CD4+CD45RA+ T cells in the development of autoimmune diabetes in the non-obese diabetic (NOD) mouse. Int Immunol 1993;5:479–489.

92 Thivolet CH, Goillot E, Bedossa P, Durand A, Bonnard M, Orgiazzi J: Insulin prevents adoptive cell transfer of diabetes in the autoimmune non-obese diabetic mouse. Diabetologia 1991;34:314–319.

93 Hutchings P, Rosen H, O'Reilly L, Simpson E, Gordon S, Cooke A: Transfer of diabetes in mice prevented by blockade of adhesion-promoting receptor on macrophages. Nature 1990;348:639–642.

94 Charlton B, Mandel TE: Progression from insulitis to beta-cell destruction in NOD mouse requires L3T4+ T-lymphocytes. Diabetes 1988;37:1108–1112.

95 Shizuru JA, Taylor-Edwards C, Banks BA, Gregory AK, Fathman CG: Immunotherapy of the nonobese diabetic mouse: Treatment with an antibody to T-helper lymphocytes. Science 1988; 240:659–662.

Jean-François Bach, MD, INSERM U 25, Hôpital Necker, 161, rue de Sèvres,
F–75743 Paris Cedex 15 (France)

Adorini L (ed): Selective Immunosuppression: Basic Concepts and Clinical Applications.
Chem Immunol. Basel, Karger, 1995, vol 60, pp 48–60

....

Selective Immunosuppression of Tumour Necrosis Factor-Alpha in Rheumatoid Arthritis

Fionula M. Brennan, Andrew P. Cope, Peter Katsikis, Deena L. Gibbons, Ravinder N. Maini, Marc Feldmann

Kennedy Institute of Rheumatology, Hammersmith, London, UK

Rheumatoid arthritis (RA) is an autoimmune disease with inflammatory manifestations in the peripheral synovial joints, which are infiltrated by activated T cells, macrophages and plasma cells. There has been much interest in the analysis of the pathogenic mechanisms which underlie this disease, with the general consensus that they are due to a complex interplay of genetic and environmental factors [1]. However, the genetic factor is not of major importance as the concordance for disease in identical twins is low, about 15% in RA in the latest series [2]. The HLA genetic component has been the studied best, although the role of cytokine and cytokine receptor genes (some of which are encoded within the MHC loci) may be of importance. Our concept of the events predisposing to autoimmune disease is illustrated in figure 1, and indicates that cytokine production is fundamental to this process. We have been particularly interested in investigating the role of cytokines in RA, and determining which of these cytokines are important in terms of disease pathogenesis. This chapter describes those studies which have indicated that tumour necrosis factor-α (TNFα) has a pivotal role in RA, and discusses therapeutic strategies which could result in diminution of the effect of this pro-inflammatory cytokine.

Cytokine Production in Rheumatoid Arthritis

The most common approach to investigate cytokine production in autoimmune disease has been to assess cytokine levels in blood and other tissue exudates. However, it was clear to us that such an approach was not sufficiently

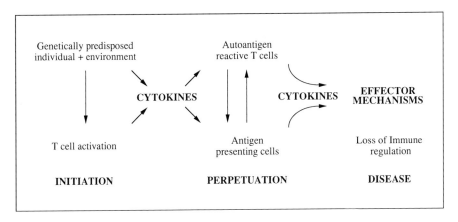

Fig. 1. Scheme of the development of autoimmunity [modified from Bottazzo et al., Lancet 1983;ii:1115].

informative. This is because cytokines in such samples are invariably biologically inactive, as they are present in an excess of cytokine inhibitors, and furthermore they do not indicate what is occurring locally, as they are at a site quite distant from the site of inflammation. Thus, we utilized an in vitro culture system using synovial membrane mononuclear cells (MNC) isolated from synovium obtained during joint replacement surgery. These cells were separated from the matrix of the synovium by enzyme digestion, and were cultured for up to 5–6 days without exogenous stimulation. Cytokine production was determined using a combination of assays to detect mRNA and protein (immunoreactive and biological active). The results are summarized in table 1 and indicate that a wide range of cytokines such as IL-1 [3], TNFα [4], granulocyte macrophage colony-stimulating factor (GM-CSF) [5], transforming growth factor-β (TGFβ) [6], IL-8 [7] and IL-10 [8] are found in abundance. Interestingly, the cytokine products of activated T cells in these cultures such as IL-2, IFNγ and LT were not detected [9, 10] whereas the corresponding mRNA had been detected (table 1). Furthermore, we showed that the presence of many of these cytokines including IL-1 [4] and GM-CSF [5] was dependent on the presence of TNFα, since its removal using neutralizing antibodies resulted in the diminution of these cytokines. We have continued these studies and have found that other cytokines (such as IL-6 and IL-8) and cell surface molecule expression are also similarly modulated on inhibition of TNFα.

Based on these studies the importance of TNFα is a pivotal cytokine in RA was proposed (fig. 2). TNFα is produced in the synovial joint principally by CD68-positive macrophages in the lining layer and the cartilage-pannus

	Cytokine	mRNA	Protein
Table 1. Summary of cytokines produced spontaneously by RA synovial cells			
	IL-1α	yes	yes
	IL-1β	yes	yes
	TNFα	yes	yes
	LT	yes	(+/–)
	IL-2	yes	(+/–)
	IL-3	no	no
	IL-4	?	no
	IFNγ	yes	(+/–)
	IFNα	yes	yes
	IL-6	yes	yes
	GM-CSF	yes	yes
	IL-8/NAP-1	yes	yes
	IL-10	yes	yes
	RANTES	yes	?
	G-CSF	yes	yes
	M-CSF	yes	yes
	TGFβ	yes	yes
	EGF	yes	yes
	TGFα	no	no
	PDGF-A	yes	yes
	PDGF-B	yes	yes

+/– = Demonstration of protein has proved difficult.

IL-1 = Interleukin 1; TNFα = tumor necrosis factor-α; LT = lymphotoxin; IFNγ = interferon-γ; IFNα = interferon-α; GM-CSF = granulocyte macrophage colony-stimulating factor; NAP-1 = neutrophil-activating peptide; RANTES = regulated on activation, normal T expressed and secreted; G-CSF = granulocyte colony-stimulating factor; M-CSF = macrophage colony-stimulating factor; TGFβ = transforming growth factor-β; EGF = epidermal growth factor; TGFα = transforming growth factor-α; PDGF = platelet-derived growth factor.

junction [11]. Furthermore, we have found using mRNA and flow cytometric techniques [12] that the p55 and p75 TNF receptors are also upregulated in RA synovium. Immunohistology confirmed these results, and indicated that the CD68-positive macrophages which produce TNFα, also express abundant TNF receptor [13].

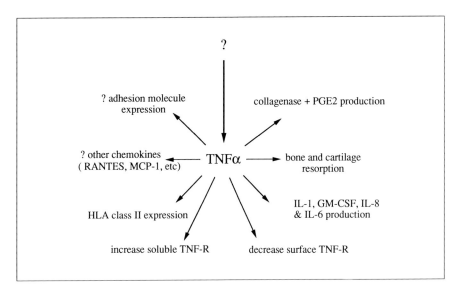

Fig. 2. Scheme summarizing effects of TNFα in RA.

Thus, in RA, there is an abundance of both TNFα (some in a bioactive form), and its cell-surface receptor. The importance of TNFα in arthritis was validated in an animal model using the collagen type II arthritic DBA/1 mice. These mice develop a chronic erosive arthritis in their hind paws typified by cartilage loss and bone erosion. If these mice were treated with an anti-TNFα antibody once the disease had started, a less severe arthritis developed, as measured by clinical, e.g. joint swelling, as well as by histological parameters such as erosions [14]. Further evidence for the importance of TNFα in arthritis is derived from human TNFα transgenic mice in which the human TNFα gene is dysregulated by replacing the 3' untranslated region of the gene with that of β-globin. These mice develop a chronic arthritis after a few months, which is totally prevented by treatment with an anti-human TNFα antibody from birth [15]. These studies taken together thus indicated that TNFα was an important molecule in the pathogenesis of RA, and helped to provide the rationale for evaluating the effect of blocking TNFα in patients with RA. This was performed in a small, open study using 20 long-standing but active RA patients that had failed many conventional drugs. A chimeric (mouse Fv/human IgG1) anti-TNFα antibody termed cA2 (produced by Centocor, Inc.) was infused using a total dose of 20 mg/kg. The drug was tolerated well by all patients and resulted in a >70% improvement in clinical parameters of disease including pain score, morning stiffness, swollen joint score and joint tenderness, as well as a fall in biochemical measurements of disease activity such as C-reactive protein, erythrocyte sedimentation rate and IL-6 levels [16]. This successful phase I/II trial has now been extended to determine if TNFα blockage can be effective in the long term. Initial results are encouraging in that all retreatment patients show improvement; however, in some patients a trend towards a shortened duration of benefit was observed. A multi-centre placebo-controlled study with 72 patients has been initiated which seeks to confirm and enlarge upon these initial findings.

Cytokine Inhibitors

An important step in understanding cytokine regulation has come from the study of specific cytokine inhibitors. These exist broadly in two groups. The major group consists of soluble cytokine receptors [17] that are derived from most of the cytokine receptors already characterised, with the exception of the G-protein-coupled chemokine receptors. The second group consists of receptor antagonists, and is represented solely by the IL-1 receptor antagonist (IL-1ra), a member of the IL-1 family which is not capable of signalling [18]. The best studied of the former group are the soluble TNF receptors (TNF-R). These were originally described in the urine of febrile patients, although both p55 and p75 TNF-R were subsequently found in biological fluids of normal volunteers including plasma and urine. We have found that both p55 and p75 TNF-R are elevated in plasma of RA patients [19], and that levels are significantly increased in synovial fluid. This suggests that in RA, the inflamed synovial joints are the source of sTNF-R. This is confirmed by the observation that both inhibitors are produced spontaneously by RA mononuclear cells cultures. These molecules are of fundamental importance since inhibition of their activity with monoclonal anti-TNF-R antibodies enhances TNF bioactivity in supernatants of these cultures [unpubl. observation], or in synovial fluids [19].

The observation that cytokine inhibitor production is augmented, although insufficiently to neutralize all TNF in RA synovium, has led us to investigate the mechanisms of sTNF-R release from monocytes. These studies confirmed that activators of monocytes including LPS, IL-1 and TNFα itself [20] caused downregulation of surface TNF receptor expression whilst inducing the production of sTNF-R in the culture supernatants.

We were particularly interested to investigate to what extent immunoregulatory cytokines such as TGFβ, IL-4 and IL-10 modulate TNF-R expression. All three cytokines inhibit TNFα and IL-1 production on monocytes [21–23] and all stimulate the secretion of the IL-1ra [24–26]. We confirmed that exposure of monocytes to IL-1α in the presence of TGFβ, IL-4 or IL-10 inhibited

TNFα production, and all three cytokines induced downregulation of surface TNF receptor expression although IL-4 and TGFβ suppressed the release of soluble TNF-R [20]. In contrast, IL-10 induced p55 and p75 sTNF-R release in a dose-dependent manner and induced TNF-R mRNA expression. These results indicate that IL-10 reduces the pro-inflammatory potential of TNF in three ways: by downregulating surface TNF-R expression whilst increasing production of soluble TNF-R and inhibiting the release of the ligand.

Immunoregulatory Cytokines in Rheumatoid Arthritis

Interleukin-10

As discussed above, in rheumatoid inflammation, despite upregulation of pro-inflammatory cytokines and the consequent production and/or activation of proteolytic enzymes, there is also an increase in homeostatic mechanisms which to some extent inhibit the ongoing inflammation. In the light of its immunoregulatory potential, we investigated to what degree IL-10 was produced and/or modulated by cytokine production in RA.

Firstly, it was observed that IL-10 was detected at both the mRNA and protein level in synovial membranes [8]. This IL-10 protein was also demonstrated by two-colour immunohistology, which indicated that both macrophages in the synovial lining layer and T lymphocytes in the mononuclear aggregates were producing IL-10. Secondly, IL-10 was produced spontaneously (at levels of up to 10 ng/ml) by RA MNC cultures for up to 5 days. Most significantly, the IL-10 was biologically active in the RA cultures since neutralisation of this IL-10 with a monoclonal antibody resulted in a 2- to 3-fold increase in the production of TNFα and IL-1 (fig. 3). This result indicates that IL-10 may be of importance in regulating the cytokine network in the RA joints. Finally, addition of exogenous IL-10 to these cultures reduced IL-1 and TNFα production two- to threefold (fig. 4), indicating that IL-10 receptors are not saturated by IL-10, and that regulation of pro-inflammatory cytokines can be achieved by further addition of IL-10. This is further strengthened by the observation [20] that IL-10 induces production of the natural inhibitors of TNF (sTNF-R), and the IL-1ra [26]. Although these results are of interest, it is not clear to what degree IL-10 given as a therapeutic modality may be of benefit in inflammatory diseases long term. Although IL-10 has predominant inhibitory effects, it is also a chemotactic factor for CD8 T cells [27], a growth factor for B cells [28], enhances antibody-dependent cellular cytotoxicity [29] and may precipitate diabetes mellitus in mice [30]. Thus, animal studies are currently being done to establish whether IL-10 treatment in vivo may modulate arthritis.

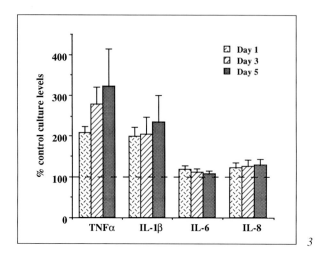

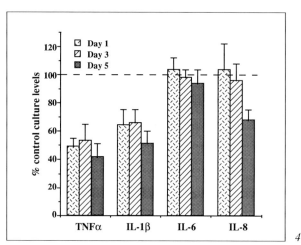

Fig. 3. Effect of neutralization of endogenous IL-10 on cytokine production in RA synovial mononuclear cell cultures. Reproduced with permission from Katsikis et al. [8].

Fig. 4. Effect of exogenous IL-10 (10 U/ml) on cytokine production in RA synovial mononuclear cell cultures. Reproduced with permission from Katsikis et al. [8].

Interleukin-4

IL-4, like IL-10, is also a cytokine with anti-inflammatory properties which could be of therapeutic benefit in arthritis. Thus, IL-4 inhibits LPS-induced IL-1, TNFα, IL-6 and PGE$_2$ production in monocytes [22, 31, 32]. In addition, it inhibits metalloproteinase biosynthesis in human alveolar macro-

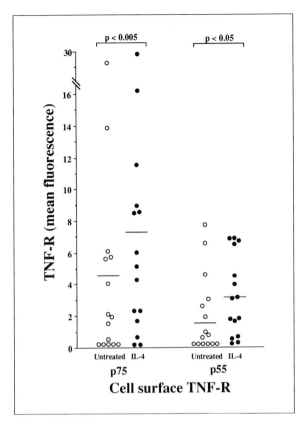

Fig. 5. Effect of exogenous IL-4 (10 ng/ml) on surface TNF-R expression in RA synovial mononuclear cell cultures as measured using anti-p55 and anti-p75 monoclonal antibodies and analysed by flow cytometry. Reproduced with permission from Cope et al. [35].

phages [33], but unlike IL-10 does not induce the tissue inhibitor of metallo-proteinases (TIMP) [Dayer, pers. commun.]. In one report using RA synovial explant cultures, IL-4 was shown to inhibit IL-1β, TNFα and IL-6 production [34]. Using our RA MNC suspension culture, we were unable to show significant inhibition of TNF bioactivity by IL-4 [35]. Suppression of LPS-induced PBMC monokine production by IL-4 does depend on pretreatment of these cells with IL-4. Thus, the inability of IL-4 to inhibit monokine production in our culture system may be due to the chronic activation state of the macrophages. Furthermore, we observed that addition of IL-4 to RA MNC cultures enhanced p55 and p75 surface TNF-R expression (fig. 5). Thus, in inflammatory sites, although IL-4 may reduce TNF bioactivity, it could enhance TNF signal transduction by increasing TNF-R surface expression. This observation taken together with the stimulatory (as well as inhibitory) effects of IL-4, and as such its role as a growth factor for T and B cells, raises the question as to how effective IL-4 will be as an immunosuppressive in RA. The outcome of projected clinical trials is awaited with great interest.

Transforming Growth Factor-Beta

TGFβ is another immunosuppressive cytokine, which like IL-4 also has stimulatory potential. Thus, TGFβ can inhibit a number of immune functions including T cell and B cell proliferation [36, 37], immunoglobulin production, natural killer (NK) and lymphokine-activated killer (LAK) cell generation [38]. In addition to its immunosuppressive role, TGFβ1 also inhibits the activity of non-haemopoietic cells such as adhesion of blood cells to endothelium [39], and osteoclast-mediated bone resorption [40]. In contrast, TGFβ1 also has stimulatory effects on connective tissue cells and in vivo stimulates the accumulation of neutrophils, macrophages and fibroblasts at the sites of wound healing [41].

Since many of the immune functions influenced by TGFβ are involved in the sequence of events occurring in diseases such as RA, interest has been directed at the possibility of using TGFβ as an immunosuppressive agent in this condition. However, despite ongoing inflammatory reactions in the synovial joint, abundant active TGFβ is detected in the synovial fluid of RA patients, and is produced spontaneously by synovial cells in culture [6]. Furthermore, the addition of exogenous TGFβ1 does not inhibit the spontaneous production of cytokine (TNFα and IL-1) by these synovial cells in culture, although, such cytokine production is inhibited by TGFβ1 in stimulated peripheral blood mononuclear cells [21].

Recently, there have been a number of in vivo studies to assess the immunosuppressive potential of TGFβ, which have yielded conflicting results. Local injection of TGFβ into rat synovial joints resulted in a rapid leucocyte infiltration with synovial hyperplasia leading to synovitis [42, 43], whereas, if injected systemically, TGFβ was found to antagonize the development of polyarthritis in susceptible rodent models [44, 45]. It is not clear in these animal models why in one circumstance TGFβ1 may actually contribute to the inflammatory process whereas in the other it inhibits the development of arthritis. Similarly, in RA it is not clear whether TGFβ is contributing to or has already partially inhibited the pathogenic events occurring within the synovial joint. It is likely that the overall effect of TGFβ1 will depend on a number of factors including the stage of activation/differentiation of the target cell, interactions with other growth factors in the environment, and the distribution and functional status of TGFβ receptors on target cells [46].

Chronic Exposure of T Cells to Tumour Necrosis Factor

The studies described above indicate that in the RA synovium, the production of pro-inflammatory cytokines (and their inhibitors) is upregulated in an environment in which the specific cytokine receptors are also increased on cells. One particularly interesting observation is that the T lymphocytes in the joint express abundant surface p75 TNF-R. We therefore investigated whether chronic exposure of activated T lymphocytes to TNF alters their function. Pretreatment of tetanus toxoid specific T cell clones with TNF for 16 days resulted in an impaired response to TT, but not IL-2 or PHA [47]. Furthermore, chronic exposure to TNF resulted in impaired cytokine production in these T-cell clones. The immunosuppressive effect of TNF on T cells in vivo was further studied using PBM from RA patients before and after treatment with a chimeric anti-TNF antibody cA2 as part of a clinical trial [16]. Treatment with anti-TNF restored the diminished proliferative response of these cells to both recall antigens and mitogens, indicating that persistent expression of TNF in vitro and in vivo impairs cell-mediated immune responses.

Concluding Remarks

This chapter has summarised our work over the last 8 years investigating to what extent cytokines contribute to the pathogenesis of RA. The identification that TNFα is a pivotal molecule in this process was an important observation in leading research from the laboratory towards the clinic and in the gradual unravelment of the inhibitory mechanisms involved. Furthermore, it has helped us to understand how the body attempts to combat inflammation. What is not clear from these studies is to what degree the pro- versus anti-inflammatory cytokines modulate the progression of disease. RA, like many other autoimmune diseases, is a chronic condition in which the disease manifestations 'wax and wane' over a period of many years. It is tempting to speculate that remission may be associated with an overall increase in cytokine inhibitory mechanisms which, albeit temporarily, inhibit inflammation. Despite the chronicity – or indeed treatment of these patients with a wide range of disease modifying drugs – the disease itself continues to progress. Further studies to learn how the production of immunosuppressive cytokines could be upregulated 'long term' at sites of inflammation is now a major goal.

References

1 Feldmann M: Molecular mechanisms involved in human autoimmune diseases: Relevance of chronic antigen presentation, class II expression and cytokine production. Immunology 1989;2: 66–71.
2 Wordsworth BP, Bell JI: Polygenic susceptibility in rheumatoid arthritis. Ann Rheum Dis 1991;50:343–346.
3 Buchan G, Barrett K, Turner M, Chantry D, Maini RN, Feldmann M: Interleukin-1 and tumor necrosis factor mRNA expression in rheumatoid arthritis: Prolonged production of IL-1α. Clin Exp Immunol 1988;73:449–455.
4 Brennan FM, Chantry D, Jackson A, Maini R, Feldmann M: Inhibitory effect of TNFα antibodies on synovial cell interleukin-1 production in rheumatoid arthritis. Lancet 1989;ii:244–247.
5 Haworth C, Brennan FM, Chantry D, Turner M, Maini RN, Feldmann M: Expression of granulocyte-macrophage colony-stimulating factor in rheumatoid arthritis: Regulation by tumor necrosis factor-α. Eur J Immunol 1991;21:2575–2579.
6 Brennan FM, Chantry D, Turner M, Foxwell B, Maini R, Feldmann M: Detection of transforming growth factor-beta in rheumatoid arthritis synovial tissue: Lack of effect on spontaneous cytokine production in joint cell cultures. Clin Exp Immunol 1990;81:278–285.
7 Brennan FM, Zachariae CO, Chantry D, Larsen CG, Turner M, Maini RN, Matsushima K, Feldmann M: Detection of interleukin 8 biological activity in synovial fluids from patients with rheumatoid arthritis and production of interleukin 8 mRNA by isolated synovial cells. Eur J Immunol 1990;20:2141–2144.
8 Katsikis PD, Chu CQ, Brennan FM, Maini RN, Feldmann M: Immunoregulatory role of interleukin 10 (IL-10) in rheumatoid arthritis. J Exp Med 1994;in press.
9 Buchan G, Barrett K, Fujita T, Taniguchi T, Maini R, Feldmann M: Detection of activated T cell products in the rheumatoid joint using cDNA probes to interleukin-2 (IL-2), IL-2 receptor and IFN-γ. Clin Exp Immunol 1988;71:295–301.
10 Brennan FM, Chantry D, Jackson AM, Maini RM, Feldmann M: Cytokine production in culture by cells isolated from the synovial membrane. J Autoimmun 1989;2:177–186.
11 Chu CQ, Field FM, Feldmann M, Maini RN: Localization of tumor necrosis factor α in synovial tissues and at the cartilage-pannus junction in patients with rheumatoid arthritis. Arthritis Rheum 1991;34:1125–1132.
12 Brennan FM, Gibbons DL, Mitchell T, Cope AP, Maini RN, Feldmann M: Enhanced expression of tumor necrosis factor receptor mRNA and protein in mononuclear cells isolated from rheumatoid arthritis synovial joints. Eur J Immunol 1992;22:1907–1912.
13 Deleuran BW, Chu CQ, Field M, Brennan FM, Mitchell T, Feldmann M, Maini RN: Localization of tumor necrosis factor receptors in the synovial tissue and cartilage-pannus junction in patients with rheumatoid arthritis. Implications for local actions of tumor necrosis factor α. Arthritis Rheum 1992;35:1170–1178.
14 Williams RO, Feldmann M, Maini RN: Anti-tumor necrosis factor ameliorates joint disease in murine collagen-induced arthritis. Proc Nat Acad Sci USA 1992;89:9784–9788.
15 Keffer J, Probert L, Cazlaris H, Georgopoulous S, Kaslaris E, Kioussis D, Kollias G: Transgenic mice expressing human tumor necrosis factor: A predictive genetic model of arthritis. EMBO J 1991;13:4025–4031.
16 Elliott MJ, Maini RN, Feldmann M, Long-Fox A, Charles P, Katsikis P, Brennan FM, Walker J, Bijl H, Ghrayeb J, Woody J: Treatment of rheumatoid arthritis with chimeric monoclonal antibodies to TNFα. Arthritis Rheum 1993;36:1681–1690.
17 Fernandez-Botran R: Soluble cytokine receptors: Their role in immunoregulation. FASEB J 1991;5:2567–2574.
18 Arend W: Interleukin 1 receptor antagonist. A new member of the interleukin 1 family. J Clin Invest 1991;5:1445–1451.
19 Cope AP, Aderka D, Doherty M, Engelmann H, Gibbons D, Jones AC, Brennan FM, Maini RN, Wallach D, Feldmann M: Increased levels of soluble tumor necrosis factor receptors in the sera and synovial fluid of patients with rheumatic diseases. Arthritis Rheum 1992;35:1160–1169.

20 Joyce DA, Gibbons D, Green P, Feldmann M, Brennan FM: Two inhibitors of pro-inflammatory cytokine release, IL-10 and IL-4, have contrasting effects on release of soluble p75 TNF receptor by cultured monocytes. 1994;submitted.

21 Chantry D, Turner M, Abney E, Feldmann M: Modulation of cytokine production by transforming growth factor-β. J Immunol 1989;142:4295–4300.

22 Essner R, Rhoades K, McBride WH, Morton DL, Economou J: IL-4 down-regulates IL-1 and TNF gene expression in human monocytes. J Immunol 1989;142:3857–3861.

23 de Waal-Malefyt R, Abrams J, Bennett B, Figdor CG, de Vries JE: Interleukin 10 (IL-10) inhibits cytokine synthesis by human monocytes: an autoregulatory role by IL-10 produced by monocytes. J Exp Med 1991;174:1209.

24 Turner M, Chantry D, Katsikis P, Berger A, Brennan FM, Feldmann M: Induction of the interleukin 1 receptor antagonist protein by transforming growth factor-β. Eur J Immunol 1991;21: 1635–1639.

25 Wong HL, Costa GL, Lotze MT, Wahl SM: Interleukin (IL) 4 differentially regulates monocyte IL-1 family gene expression and synthesis in vitro and in vivo. J Exp Med 1993;177:775–781.

26 de Waal-Malefyt R, Yssel H, Roncarolo MG, Spits H, de Vries JE: Interleukin-10. Curr Opin Immunol 1992;4:314.

27 Jinquan T, Larsen CG, Gesser B, Matsushima K, Thestrup-Pedersen K: Human IL-10 is a chemoattractant for CD8+ T lymphocytes and an inhibitor of IL-8 induced CD4+ T lymphocyte migration. J Immunol 1993;151:4545–51.

28 Ishida H, Hastings R, Kearney J, Howard M: Continuous anti-interleukin 10 antibody administration depletes mice of Ly-1 B cells but not convential B cells. J Exp Med 1992;175:1213–1220.

29 Te Velde AA, de Waal Malefyt R, Huijbens RJF, de Vries JE, Figdor CG: IL-10 stimulates monocyte FcγR surface expression and cytotoxic activity. Distinct regulation of antibody-dependent cellular cytotoxicity by IFN-γ, IL-4 and IL-10. J Immunol 1992;149:4048.

30 Wogensen L, Huang X, Sarvetnick N: Leukocyte extravasation into the pancreatic tissue in transgenic mice expressing interleukin 10 in the islets of Langerhans. J Exp Med 1993;178: 175.

31 Hart PH, Vitti GF, Burgess DR, Whitty GA, Piccoli DS, Hamilton JH: Potential antiinflammatory effects of interleukin-4: Suppression of human monocyte tumour necrosis factor α, interleukin-1, and prostaglandin F2. Proc Natl Acad Sci USA 1989;86:3803–3807.

32 Te Velde AA, Huijbens K, Heije JE: Interleukin-4 (IL-4) inhibits secretion of IL-1β, tumour necrosis factor α, and IL-6 by human monocytes. Blood 1987;76:1392–1397.

33 Lacraz S, Nicod I, Galve-de Rochemonteux B, Baumberger C, Dayer JM, Weleus HG: Suppression of metalloproteinase biosynthesis in human alveolar macrophages by interleukin-4. J Clin Invest 1992;90:382–386.

34 Miossec P, Briolay J, Dechanet J, Wijdenes J, Martinez-Valdez H, Banchereau J: The inhibition of the production of pro-inflammatory cytokines and immunoglobulins by interleukin-4 in an ex vivo model of rheumatoid synovitis. Arthritis Rheum 1992;35:874–883.

35 Cope AP, Gibbons DL, Aderka D, Foxwell BMJ, Wallach D, Maini RN, Feldmann M, Brennan FM: Differential regulation of tumour necrosis factor receptors (TNF-Rs) by IL-4: Upregulation of p55 and p75 TNF-Rs on synovial joint mononuclear cells. Cytokine 1993;5:205–212.

36 Kehrl JJ, Wakefield LM, Roberts A, Jakowlew S, Alvarez-Mon M, Derynck R, Sporn MB, Fauci AS: Production of transforming growth factor-β by human T lymphocytes and its potential role in the regulation of T cell growth. J Exp Med 1986;163:1037–1050.

37 Kehrl JJ, Roberts AB, Wakefield LM, Jakowlew S, Sporn MB, Fauci AS: Transforming growth factor β is an important immunomodulatory protein for human B lymphocytes. J Immunol 1986;137:3855–3860.

38 Ranges GE, Figari IS, Espevik T, Palladino MA: Inhibition of cytotoxic T cell development by transforming growth factor beta, reversal by recombinant tumour necrosis factor. J Exp Med 1987;166:991–998.

39 Gamble JR, Vadas M: Endothelial adhesiveness for blood neutrophils is inhibited by transforming growth factor-β. Science 1988;242:97–99.

40   Tashjian AH, Voelkel EF, Lazzaro M: Alpha and beta transforming growth factors stimulate prostaglandin production and bone resorption in cultured mouse calvaria. Proc Natl Acad Sci USA 1985;82:4535–4538.

41   Massague J: The TGFβ1 family of growth and differentiation factors. Cell 1987;49:437–438.

42   Allen JB, Manthey CL, Hand AR, Ohura K, Ellingsworth L, Wahl SM: Rapid onset synovial inflammation and hyperplasia induced by transforming growth factor β. J Exp Med 1990;171: 231.

43   Fava RA, Olsen NJ, Postlethwaite AE, Broadly KN, Davidson JN, Nanney LB, Lucas C, Townes AS: Transforming growth factor β1 (TGF-β1) induced neutrophil recruitment to synovial tissues: Implication for TGF-β- driven synovial inflammation and hyperplasia. J Exp Med 1991;173:1121.

44   Brandes ME, Allen JB, Ogawa Y, Wahl SM: Transforming growth factor β1 suppresses acute and chronic arthritis in experimental animals. J Clin Invest 1991;87:1108.

45   Kuruvilla AP, Shah R, Hochwald GM, Liggit HD, Palladino MA, Thorbecke GJ: Protective effect of transforming growth factor β1 on experimental autoimmune diseases in mice. Proc Natl Acad Sci USA 1991;88:2918.

46   Massague J: Receptors for the TGFβ1 family. Cell 1992;69:1067.

47   Cope AP, Londei M, Chu NR, Cohen SBA, Maini RN, Brennan FM, Feldmann M: Chronic exposure to TNF *in vitro* impairs the activation of T cells through the T cell receptor/CD3 complex: reversal *in vivo* by anti-TNF antibodies in patients with rheumatoid arthritis. J Clin Invest 1994;in press.

F. M. Brennan, MD, Kennedy Institute of Rheumatology, Sunley Building, 1 Lurgan Avenue, Hammersmith, London W6 8LW (UK)

Adorini L (ed): Selective Immunosuppression: Basic Concepts and Clinical Applications.
Chem Immunol. Basel, Karger, 1995, vol 60, pp 61–78

....

MHC Blocking Peptides and T-Cell Receptor Antagonists: Novel Paths to Selective Immunosuppression?

Jörg Ruppert [a], *Alessandra Franco* [a], *Jeff Alexander* [a], *Ken Snoke* [a],
Glenn Y. Ishioka [a], *Dawne M. Page* [b], *Stephen M. Hedrick* [b],
Luciano Adorini [c], *Howard M. Grey* [a], *Alessandro Sette* [a, 1]

[a] Cytel, San Diego, Calif.;
[b] Department of Biology and the Cancer Center, University of California
 at San Diego, La Jolla, Calif., USA;
[c] Roche Milano Ricerche, Milan, Italy

Recognition of peptide fragments by antigen-specific T cells is an event
crucial for the generation of specific immunity. Peptides derived from intracellular processing of protein antigens bind to major histocompatibility complex
(MHC) molecules and are finally presented on the surface of antigen-presenting cells (APC). It is the bi-molecular complex of antigenic peptide and MHC
that is actually recognized by the T-cell receptor (TCR) [1, 2]. MHC class II
molecules do not exhibit the highly selective binding requirements for their ligands typical of other membrane receptors. As a consequence, a multitude of
antigens potentially can bind to a few MHC class I or II restriction elements.
Several detailed studies of the structure-function relationship between MHC
determinants and peptides revealed the basic rules for peptide binding [3–5].
In general, peptide binding is accomplished by interaction of two or three main
anchor residues in polymorphic pockets of the MHC molecule. In the case of
class II, few restrictions regarding the length of a given antigen peptide were
found. This is in strong contrast with the strict peptide length requirements determined for MHC class I molecules [6]. On the other side of the tri-molecular

[1] We thank Joyce Joseph for her excellent assistance in preparing the manuscript.
Partially supported by National Institutes of Health Grant AI18634 (H.M.G.).

complex, the TCR recognizes a given peptide antigen only in context with a certain MHC allele and represents a highly variable and specific receptor.

Antigen-presenting cells do not exclusively present foreign antigen material. Rather, it has been shown that self peptides constitute the majority of the peptides presented. In healthy individuals, however, self peptides do not trigger self-destructive T-cell responses. This is due to several mechanisms, inducing the destruction of autoreactive T cells specific for self peptides during a selection process in the thymus [7] and the peripheral inactivation of autoreactive T cells (peripheral tolerance) [8]. Failure of these mechanisms can lead to loss of tolerance to particular self antigens and result in autoimmunity, with the immune system attacking cells or structures of the host's body.

A significant characteristic of a number of different autoimmune diseases is their strong association with certain MHC alleles [9]. Therefore, it appeared feasible to interfere with the onset of such diseases by inhibiting the T-cell populations restricted to these particular MHC haplotypes and potentially involved in disease pathogenesis.

Indeed, peptides with high affinity for a particular MHC molecule have been used, both in vivo and in vitro, to demonstrate inhibition of T-cell function through competition for antigen presentation [10–15]. In particular, promising results have been obtained by several groups in the murine autoimmune disease, experimental allergic encephalomyelitis (EAE), which represents an animal model of multiple sclerosis in humans. Here, the onset of MHC-restricted, T-cell-mediated disease was prevented by administration of high doses of MHC binding peptides [16–18]. Although it had been suggested that this inhibition was, in fact, mediated by MHC blockade, a more detailed analysis was required to critically evaluate the feasibility of this approach for treatment of autoimmune diseases. Experiments performed in our laboratory to answer this question are outlined in the following section.

Therapeutic Potential of MHC Blocking Peptides

During the course of experiments designed to identify an optimal MHC blocking peptide, we generated a non-natural peptide which possessed a strong binding capacity and high serum stability: CY-760.50. The sequence of this particular 13-mer peptide was based upon studies determining the specific motif recognized by various DR molecules [19]. The peptide contained the crucial contact residues for the MHC class II DR1 molecule incorporated into a polyalanine peptide backbone [20]. In order to increase the solubility of the peptide, a lysine residue was introduced in a position which did not contribute to MHC binding. To increase resistance of the peptide against enzymatic degradation, *d*-alanine residues were included at the N- and C-termini, and the carboxyl terminal group was amidated. The relative binding affinity of CY-760.50 to DR1 was found to be 3- to 30-fold stronger than other previously identified high-affinity peptides.

Upon testing the ability of this peptide to inhibit presentation of the hemagglutinin (HA) 307–319 peptide by a homozygous DR1 antigen-presenting cell to a specific DR1-restricted T-cell clone (Cl-1), we were not surprised to find that CY-760.50 had potent inhibitory activity. The important question to be addressed next was the longevity of this blocking effect. To our own surprise, no inhibition of antigen presentation was observed once CY-760.50 was removed from the extracellular fluid surrounding the APC [21].

It is still a matter of debate whether peptide competition for MHC molecules occurs only at the cell surface, or if peptide competition for MHC binding could also take place in endosomal compartments, possibly during the recycling of class II MHC molecules [22–24]. In our experiments, similar results were obtained with either live or glutaraldehyde-fixed APC. This result excludes the possibility that recycling and unloading of the inhibitor peptide from class II MHC molecules or the binding of stimulator HA peptide to newly synthesized DR molecules might explain the lack of long-lasting MHC blockade. Furthermore, the possibility that this effect was due to rapid dissociation of CY-760.50 from DR molecules appears unlikely, inasmuch as we observed in vitro dissociation rates for this peptide that were even slower than other high-affinity DR1 binding peptides.

To determine whether these in vitro observations could also apply to an in vivo situation, we studied the pharmacokinetics and the duration of biologic activity of CY-760.50 in C57BL/6 mice. An important step, in order to correlate pharmacokinetics and duration of action of CY-760.50 in vivo, was to establish a surrogate marker assay which was able to detect inhibition of complex formation between antigenic peptides and I-A^b molecules [21]. Here, we found that following a bolus intravenous injection, the peptide CY-760.50 was rapidly cleared from the circulation. In parallel to clearance, loss of Ag blocking capacity in peripheral blood lymphocytes (PBL) occurred within a few minutes of CY-760.50 injection. This result clearly showed that duration of action of MHC blockade after a bolus intravenous injection is directly related to the presence of an effective concentration of MHC antagonist in the fluid phase. Therefore, in a hypothetical therapeutic application of MHC blocking peptides, it appears crucial to maintain effective concentrations in the extracellular fluids throughout the treatment. This appears particularly problematic for short peptides, with usually poor pharmacokinetics and rapid clearance and/or degradation in vivo [25, 26]. Thus, it is not surprising that in vivo studies have required the administration of relatively large quantities of MHC blocking peptides, which are usually emulsified together with antigen in adjuvant, in order to demonstrate effective inhibition of a T-cell immune response [10, 11, 17, 18]. As an alternative, in some cases, repeated administration of soluble peptide [27] or sustained release by mini-osmotic pumps has also been found effective [12]. Limitations regarding pharmacokinetics and pharmacodynamics appear particularly relevant if treatment of chronic diseases such as rheumatoid arthritis or multiple sclerosis is to be attempted. In general, the therapeutic value of MHC blocking peptides in vivo strongly depends upon convincing solutions regarding drug delivery and increased potency.

TCR Antagonism: A Promising Alternative

During studies testing various peptides and analogs in parallel for MHC binding and inhibition of T-cell proliferation, we made the observation that single substitution analogs of antigenic peptides appeared to be much more effective inhibitors in cellular assays than unreleated peptides with high MHC binding capacity. Both groups of peptides bound equally well to MHC molecules, but the analogs were found to be more effective in inhibiting T-cell proliferation by approximately two orders of magnitude. Another important feature of these analogs was the observation that their effect was antigen-specific, in the sense that analogs of the HA antigen exclusively inhibited proliferation of HA-specific T cells, while analogs of the tetanus toxoid (TT) antigen only inhibited TT-specific T cells. It thus became apparent that antigen analogs could act as antagonists of the TCR [28]. A test system was established to specifically measure this TCR antagonism effect. The so-called prepulse assay followed a different protocol than the antigen competition assay. For the prepulse assay, APC were at first pulsed with a suboptimally stimulating antigen dose for 2 h, and after removal of unbound peptide, the analog was allowed to bind in a second pulse for another 2 h. Finally, free analog was removed from the cultures by washing, and the APC were co-cultured with T cells specific for the particular antigen. In the competition assay, APC were pulsed with both antigen and competitor simultaneously. Peptides unrelated to the antigen with a strong MHC binding affinity only showed inhibition in the competition assay. In contrast, only single amino acid substitution (SAAS) analogs of the antigen showed inhibition in the prepulse assay. Therefore, we concluded that the effect of these analogs had to take place on the side of the T-cell receptor, rather than just blockage of the MHC molecule.

Using this approach, Ostrov et al. [29] have analyzed two different DR5-restricted T-cell clones. One clone recognizes the native antigen HA 307–319, while the other is specific for a single amino acid-substituted analog. They were able to demonstrate that the peptide specific for one clone functioned as antagonist for the other clone and vice versa. However, the sequences of the TCR of these two clones were found to be strikingly similar [29]. This result underlined the exclusive antigen specificity of T-cell inhibition mediated by peptide analog.

The molecular mechanism underlying the phenomenon of T-cell inhibition by analogs of the wild-type antigen and its mechanism of action were investigated in more detail. First, we sought to gain more insight into the correlation of structure and function of antagonistic peptides. To determine whether the ligand-receptor interaction of the antagonist was of a competitive or noncompetitive nature, the proliferation of human Cl-1 T cells was measured in response to a dose range of the antigen HA 307–319 in the presence of increasing amounts of an HA analog (HA 307–319Q[313]). The results showed a curve characteristic for a competitive ligand. More specifically, with increasing amounts of antagonist, higher amounts of antigen were needed to reach the same maximal response. Thus, it could be concluded that the antagonist binds reversibly to the same receptor site on the TCR as the antigen, and that the tested antigen analog functions as a competitive antagonist [30].

To determine the role of different residues within the antigen peptide, we generated a large panel of single amino acid substitutions and first tested their DR1 binding capacity using a previously established protocol [31]. Using this approach, we were able to define the residues that are crucial for interaction with the MHC molecule. Next, a subset of peptides homogenous in MHC binding was selected, and the capacity to stimulate T cells was tested. Drastic effects on peptide antigenicity were observed, with most substitutions introduced at five residues within the HA peptide. These five positions within the peptide were thus identified as major T-cell contact residues [5, 31]. Testing these analogs in the aforementioned prepulse assay revealed that modification of each of the five major T-cell contact sites can generate a powerful antagonist [30].

We also analyzed the effect of sequential re-incorporation of four of the five T-cell contact residues into a poly-A peptide backbone. Poly-A peptide analogs were used by several groups as an elegant way to determine the minimum structural requirement for a particular antigen peptide function binding MHC I or II [20, 32]. In our experiments, we used analogs which were depleted of all TCR contact residues but that still contained all MHC binding sites. Comparing our data for antigenicity and antagonism revealed that the increasing similarity of the analog to the original antigen was paralleled by an increasing antagonistic potency. Re-incorporation of all five T-cell contact residues completely restored antigenicity. Our conclusion from these data was that effective T-cell activation appeared to require a certain affinity threshold. Analogs binding well to the MHC molecule but falling below this threshold level necessary to trigger T-cell signaling would then be able to act as TCR antagonists and prevent TCR-mediated stimulation. A similar phenomenon has been implemented for B-cell activation and immunoglobulin receptors [33, 34].

Clonal anergy has been implemented as another mechanism for apparent lack of T-cell proliferative response [35–41]. Inasmuch as clonal anergy or T-cell tolerance appears to be induced by inappropriate signaling or lack of a costimulatory signal [37, 38], this phenomenon might also play a role for the effect of TCR antagonism. This notion is discussed in the following section.

Antagonism versus Tolerance or Anergy

T-cell clonal anergy has been described by Schwarz [39] as a state of pro-liferative unresponsiveness resulting from the inability to produce IL-2. Ear-lier studies demonstrated that nonreponsiveness can be induced in the absence of APCs [41] in a hemagglutinin antigen-specific, DR1-restricted clone. In this system, it is likely that the antigen is 'self presented' by class II expressing T cells, but in the absence of a costimulatory signal. Following this approach, we found that overnight incubation of the same T-cell clone used for antago-nism studies with the antigen HA 307–319 in absence of a professional APC actually induced tolerance in a dose-dependent manner [30]. T cells incubated with 10–100 µg/ml of the antigen were subsequently virtually nonresponsive to APCs pulsed with a high dose of the same peptide. However, even the highest dose (100 µg/ml) of the antagonist HA-Q^{313}, either in the presence or absence of the suboptimal dose of antigen as used in prepulse assays, did not induce tolerance, and the T cells could readily be stimulated by APCs pulsed with graded doses of antigen.

In a different set of experiments, we took a more direct approach to es-tablish whether T cell tolerance was involved in the lack of proliferation ob-served with these clones. We re-isolated T cells that were co-incubated with APCs either prepulsed with a suboptimal amount of antigen followed by dif-ferent doses of antagonist, pulsed with the antagonist alone, or left unpulsed. When these T cells were isolated from cultures after 20 h and stimulated with APC presenting the antigen, they were still able to respond normally. In con-trol experiments, proliferation was effectively inhibited when the T cells were left in culture with prepulsed APC presenting the antagonist together with antigen. Therefore, we concluded that the mechanisms leading to TCR antago-nism must be different from those leading to T-cell anergy. Data reported re-cently for the mouse system suggest that a nonantigen peptide can induce tol-erance [42]. Therefore, it appears that the effects of TCR antagonism and

T-cell anergy can overlap in some systems. In our experiments, however, T-cell anergy did not serve as an explanation for the mechanism of TCR antagonism.

In the following section, we outline experiments leading to the proposal of a molecular model for the function of TCR antagonists.

TCR Antagonism: Suggestion of a Model

The observations described above suggested that affinity plays an important role in determining the functional outcome of peptide-TCR interactions. To better understand the mechanism involved in TCR antagonism, we analyzed the effect of an antagonistic antigen analog on different biochemical events of the T-cell activation cascade [43]. Utilizing the DR1-restricted HA antigen in the prepulse assay system, we found late events of T-cell activation such as expression of the IL-2 receptor to be inhibited by the antagonist in a dose-dependent fashion. Along with this late event of the activation cascade, earlier events, such as IP formation and CA^{2+} flux, were also inhibited in an antagonist dose-dependent fashion. However, we found that the initial contact between APC and T cells leading to the formation of hetero-conjugates was still intact, even in the presence of a high amount of antagonist. Thus, a suboptimal antigen dose was sufficient to induce APC-T-cell binding and could not be impaired by pulsing the same APC with up to a 1,000-fold excess of the antagonist.

Moreover, presentation of antigen and antagonist separately on different APCs did not have any effect on antigen-induced T-cell activation. In control experiments, we established that no conjugate formation was induced by the antagonist itself in the absence of antigen. Thus, it was obvious that presence of antigen and antagonist together on the same APC was a prerequisite for TCR antagonism to occur. More specifically, a suboptimal antigen dose was sufficient to engage T cells leading to proliferation. Only the minimum of a 3- to 10-fold excess of antagonist was needed to effectively block antigen-induced T-cell activation [28, 30]. These observations further supported the conclusion that a mechanism more complex than simple MHC blockade or receptor saturation of the TCR had to be considered. Recent studies by other groups [44, 45] have already implied that TCR-mediated signaling can be rather differential and does not necessarily induce the entire intracellular activation cascade.

Formation of conjugates between APC and T cells can be referred to as 'vertical signaling', and might also include events such as upregulation of different adhesion molecules [42]. On the other hand, 'horizontal signaling' can be viewed as a cascade of intracellular events ultimately leading to T-cell proliferation.

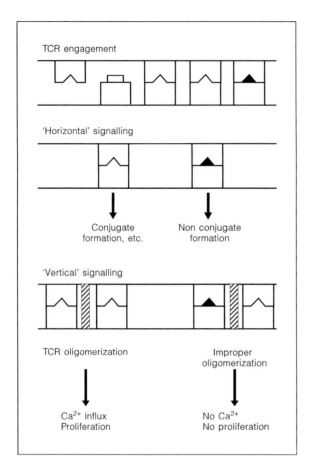

Fig. 1. A model of TCR antagonism.

TCR antagonists apparently block intracellular signaling but not cell-cell interaction. On the basis of these considerations, we proposed a model to explain the molecular mechanism of TCR antagonist function (fig. 1). During T-cell activation mediated by antigen bound to the MHC, a certain threshold level of affinity for the TCR might be required to initiate cell-cell interaction, starting with only a few surface receptor molecules. This event is followed by microclustering of a low but distinct number of bound receptors required for 'vertical signaling', as also proposed by others [46]. However, in the case of an APC presenting a mixture of antigen and antagonist, the presence of a suboptimal amount of antigen might be sufficient to bind a limited number of receptors, induce conjugate formation, and trigger 'horizontal signaling', but the presence of bound antagonist might eventually lead to mixed receptor oligomers which are somehow incapable of triggering intracellular activation events.

Oligomer formation being involved in effective signaling is also implied by recent studies of the three-dimensional structure of the DR1 molecule [46]. These studies suggest the possibility that formation of dimers of DR1 detected in the structure of DR1 crystals could be reflective of dimerization of the TCR on the engaged T cell while binding to MHC-antigen complexes.

Our model of mixed and thereby nonfunctional oligomers finds its parallels in other systems. For example, it has been demonstrated that inhibition of signal transduction was impaired by a mixed trimer of the p53 molecule [47, 48]. Our hypothetical model would also readily explain antagonist function found in the MHC class I system on CD8+ cells, showing a basically similar antagonistic effect as reported by us for CD4+ cells [49]. Furthermore, Allen and co-workers have shown that antigen analogs are also capable of antagonizing superantigen-induced activation of Th1 and Th2 T cell clones and hybridomas [Evavold, Sloan-Lancaster, Allen PM, unpubl. information]. Recent data reported by Germain's group [50] added further evidence to the phenomenon of TCR antagonism by showing that differences in T-cell-receptor-mediated signaling of an MHC class II-restricted T-cell clone could be attributed to MHC class II-peptide complexes with mixed agonist/antagonist properties [50]. Therefore, TCR antagonism appears to be considered a rather general phenomenon, and its study might yield valuable information in many different experimental systems.

TCR Antagonist Peptides Inhibit Different T-Cell Lines

TCR antagonism has potential also to be utilized as a highly specific and powerful means for the treatment of allergic and autoimmune diseases. Several reports demonstrated that the repertoire of T cells involved in experimentally induced autoimmune diseases can be very heterogeneous in mice [51], and possibly also in humans [52], while others reported a rather restricted response [53–55]. Investigations of TCR heterogeneity in multiple sclerosis [56, 57] and rheumatoid arthritis [58, 59] have led to somewhat contradicting observations. Because of these uncertainties, for the treatment of autoimmune diseases or allergies using antigen antagonists it would be very desirable to demonstrate that this peptides are able to inhibit most or all T-cell clones specific for the target epitope.

It has been known for several peptides that only one residue can be crucial for a specific T-cell response mediated by different T-cell clones. One example was described for the OVA 323–329 peptide, where a variety of T-cell clones were analyzed and no longer responded to this antigen when the crucial residue H331 was substituted.

Experiments performed in our laboratory showed a similar phenomenon for an analog of the HA 307–319 antigen [60]. More specifically, we tested the ability of TCR antagonist peptides to inhibit a panel of five different DR4w4-restricted, influenza hemagglutinin 307–319-specific T-cell lines. Each of these lines expressed a distinctly different TCR gene arrangement, as determined by PCR analysis of their Vβ and Jβ usage. Analysis of the antigen fine specificity of these clones was determined by measuring their proliferative response to a panel of single amino acid substituted analogs. No substitutions were made at residues crucial for MHC binding. Monitoring lack of T-cell proliferation, we were able to detect the most important residues for activation of the five T-cell lines tested. Residues for which most of the analogs lost capacity to stimulate a response were regarded as T-cell contact residues for the respective T-cell line. The analogs identified by this approach were then tested in the prepulse assay to assess their potency as TCR antagonists. We identified a set of four non-stimulatory analogs, each able to antagonize one or more T-cell line. One analog carrying an alanine instead of lysine in position 316 was found to be a potent TCR antagonist for four of five lines.

In a different approach, T-cell contact residues were sequentially reintroduced into a polyalanine backbone peptide and tested for T-cell stimulation and antagonism. Interestingly, one of these simple synthetic peptides was also found to antagonize four of the five lines, reaching 50% inhibition (IC_{50}) with a dose range of 50–100 µg/ml. Therefore, the poly-A analog was found somewhat less effective as compared to the SAAS analog of the original antigen, with an IC_{50} of 1–50 µg/ml, depending on the particular clone tested [60]. Interestingly and importantly, the one T-cell line out of five that was escaping TCR antagonism was a different one for the antigen analog and the poly-A analog. Thus, it can be excluded that, in this instance, certain T-cell lines remain unaffected by any antagonist. It is important to note that it is possible to generate analogs that are capable of antagonizing Vβ3- and Vβ5-carrying T cells. The two most potent analogs described above were also found to inhibit the response of a DR1-restricted T-cell clone expressing the Vβ13 TCR gene. Therefore, it appears that TCR antagonists are capable of inhibiting multiple T-cell clones with different allelic restrictions and expressing different Vβ segments specific for the same antigen. These findings raise some optimism on the feasibility of TCR antagonists as a therapeutic strategy. The following section describes experiments using TCR antagonists as inhibitors of EAE in mice.

EAE: In vivo Application of TCR Antagonists

In SJL (H-2^s) mice, a well-defined antigen, the proteolipid protein (PLP) 139–151 epitope, can be utilized to induce EAE. Previous reports demonstrated that blockade of the MHC by peptides that are unrelated to the antigen but bind strongly to the MHC molecules associated with the disease (I-A^s) might be an effective treatment [17]. Independently, analogs of the myelin basic protein (MBP) were shown to inhibit EAE effectively in H-2^u mice [16, 61]. In subsequent experiments, it was shown by McDevitt and co-workers [61] that it is possible to inhibit EAE by pretreatment with an analog of the encephalitogenic antigen, followed 1–2 weeks later by immunization with the antigen.

As observed by others, the pathogenesis of EAE involves a heterogenous population of T cells [51]. Therefore, a successful therapeutic compound should preferentially be effective against a variety of T-cell populations that share the same antigenic specificity. The results described in the previous section encouraged us to perform in vivo studies to further address the feasibility of TCR antagonism as a therapeutic strategy.

Within the PLP 139–151 peptide, we detected significant effects on MHC binding and/or T-cell recognition when different amino acid substitutions were introduced at positions 142–148 (wild-type peptide sequence GKWLGHP), including at least one conservative and one nonconservative substitution for each position. Apparently, the N- and C-terminal parts of the peptide did not contribute significantly to MHC or TCR interaction, since substitutions in these positions did not show any significant effect. Binding to the I-A^s molecule was impaired by substituting positions 145 and 148, and these residues were therefore regarded as the MHC anchor residues within the PLP peptide.

In order to identify potentially antagonistic analogs of the PLP 139–151 peptide, synthetic analogs with good binding affinity to the I-A^s MHC molecule were first tested for their T-cell stimulatory capacity. One residue, W144, was found to be the most critical TCR contact residue in this antigen, because substitutions with amino acid A or K in this position abolished proliferation of all seven independently derived PLP 139–151-specific T-cell clones tested. Furthermore, five of seven clones tested failed to respond when an F was substituted in position 144. Substitutions in a variety of other positions had different effects, depending on the T-cell clone used, and appeared to be less crucial for TCR recognition. For example, some substitutions at positions 142 and 146 also affected T-cell recognition for most, and in one case (K142), all T-cell clones, but substitution with other residues yielded a strongly stimulatory peptide antigen.

The phenomenon of one amino acid dominating the interaction between MHC-antigen complex and TCR as observed with the W144 residue in PLP 139–151 finds its parallels in other systems, such as in the OVA 323–339, where H331, in the pigeon cytochrome C system, where K99, and in the influenza hemagglutinin system, where K311 and K316 play a crucial role for TCR recognition [5, 30, 62]. Ehrich et al. [63] recently also described a dominating effect of one or a few charged or aromatic residues on the interaction between the MHC-antigen complex and TCR.

When these nonstimulatory analogs were tested for TCR antagonism in a prepulse assay, the strongest inhibition of T-cell proliferation, i.e. TCR antagonist activity, was detected with analogs with A, K and F substitutions in position 144, the major TCR contact site in the PLP peptide. These antagonists were able to reach a 50% inhibitory dose for various T-cell clones at concentrations in the range from 0.7 to 8 μM. Certain substitutions at other positions, such as A142, F143, A145, K or F146, and A147 had significantly higher IC_{50} doses (10–50 μM). Therefore, we focused on position 144 and synthesized six additional analogs. Of these, the residue L in position 144 inhibited three clones at a concentration of 10 μM or less.

Inasmuch as no single analog was capable of functioning as an antagonist and inhibiting proliferation of all T-cell clones studied, we combined two different analogs carrying either Y or L in position 144 in a pool to be used in vivo. These two analogs together were able to inhibit five of six T-cell clones in vitro.

The failure to identify one analog able to inhibit all T-cell clones tested is in line with other reports by Snoke and co-workers for the human DR system (reviewed herein) and by Jameson et al. [49], who reported similar observations in the mouse MHC class I system.

In our first in vivo approach, we evaluated the effect of this pool of antagonistic analogs on EAE induction when injected simultaneously with the encephalitogenic PLP peptide in a complete Freund's adjuvant emulsion. As a control, the ROIV peptide was included [64]. This peptide was found to be an effective inhibitor in vivo of EAE. It is unrelated to PLP and exhibits a high binding affinity for I-A^s MHC molecules.

The results obtained in these experiments clearly showed that a pool of two analogs was superior (92% inhibition) to the MHC blocker ROIV and to the injection of single analogs in inhibiting induction of EAE as monitored over a time period of 35 days. The additive effect of the two antagonist peptides strongly suggested that each peptide was capable of inhibiting a limited array of antigen-specific T-cell clones, while the two analogs combined affected almost all clones involved. When the excess dose of the antagonist pool was varied, we still observed approximately 50% inhibition of EAE induction with equimolar amounts of antigen and analogs. By contrast, a 10-fold molar excess of the MHC blocking peptide ROIV was required to reach a similar level of inhibition in vivo.

These data give an impressive view of the potential of TCR antagonists as therapeutic agents in vivo, and correlate with previous studies in which antigen analogs were demonstrated to be highly effective in vivo inhibitors of antigenic responses [16, 61, 65].

In another series of experiments, we administered the TCR antagonist pool prior to injection of the antigen PLP. In parallel, we pretreated mice with a high dose of the wild-type antigen in incomplete Freund's adjuvant or the ROIV peptide. To no surprise, pre-injection with the PLP antigen resulted in complete inhibition of EAE induction. This phenomenon has previously been observed in other systems [61, 65–68], and is most likely due to the induction of T-cell tolerance. However, the TCR antagonist pool was also found to inhibit EAE induction, albeit to a lower degree (50%), suggesting induction of a prolonged T-cell hyporesponsiveness in vivo. The MHC-blocking peptide ROIV was completely ineffective, confirming our observations for in vivo effects of MHC blockers, as outlined in a previous section of this review [21].

It is important to note that the encephalitogenic antigens PLP and MBP of the mouse disease model EAE are likely to also play a role as autoantigens in the pathogenesis of multiple sclerosis in humans. Both proteins are also recognized by human autoreactive T cells, and mapping of these peptides is underway in several laboratories [69–71]. Taken together, the results of these efforts might, in the not too distant future, lead to testing of antigen-specific strategies that allow successful immunomodulation of autoimmune diseases.

TCR Antagonists as a Tool to Study Negative Selection

Finally, TCR antagonists proved to be helpful to study cellular signaling requirements involved in T-cell maturation in the thymus. Cells entering the thymus at an early stage in their development do not express the TCR molecule complex, nor do they carry the CD4 or CD8 accessory molecules [7], and they are therefore disignated 'double negative'. Within the thymus, T cells pass a stage where they become CD4+/CD8+ double-positive, and start to express TCR. They ultimately develop into CD4- or CD8-positive cells, depending on the specificity of their TCR for MHC class II or class I molecules, respectively (positive selection). Another crucial event of thymocyte development at this stage is the elimination of potentially autoreactive T cells specific for self antigens (negative selection). The cellular and signaling requirements for negative or positive selection are still relatively obscure.

We explored the effect of antigen analogs on T-cell maturation and thymocyte-negative selection [72]. For these experiments, we used mice that are transgenic for a TCR specific for moth cytochrome C or pigeon cytochrome C peptides bound to H-2E^k class II molecules as a source of antigen-specific T cells and thymocytes. Negative selection can be analyzed in vitro by incubation of thymocytes with defined APC populations. Using this in vitro system, we found that several nonstimulatory TCR peptide antagonists can also mediate deletion, or negative selection, of double-positive thymocytes. These findings are in good agreement with data reported previously showing that negative selection can occur in response to sequence-substituted peptides that are not capable of inducing T-cell activation [73], or that lower amounts of peptide are needed to induce negative selection than are required for minimal T-cell activation [74, 75].

TCR antagonists are now being used by different groups to unravel the role peptides play in negative and positive selection in the thymus [76–78].

Conclusion

Taken together, the data presented herein suggest that antigen analogs may serve as powerful tools in many areas of immunological research and may help delineate the mechanisms involved in T-cell maturation and differentiation, and in T-cell activation and/or signaling events.

Finally, TCR antagonists might also have therapeutic potential for use in allergies and autoimmune diseases. In order for their value to be realized, however, a dramatic increase in knowledge about the exact molecular nature of antigenic peptides involved in such diseases is required. New technologies allowing elution and characterization of naturally processed antigens from MHC class I and class II molecules may help this field progress and allow generation of compounds interfering with activation of pathogenic autoaggressive T cells.

References

1   Davis MM, Chien YH: Topology and affinity of T cell receptor mediated recognition of peptide-MHC complexes. Curr Opin Immunol 1993;5:45–49.
2   Sette A, O'Sullivan D, Krieger J, Karr RW, Lamont AG, Grey HM: MHC-T cell interactions; An overview; in Gefter ML (ed): Seminars in Immunology. Philadelphia, Saunders, 1991, vol 3, pp 195–202.
3   Sette S, Sidney J, Oseroff C, Del Guercio MF, Southwood S, Arrhenius T, Powell MF, Colón SM, Gaeta FCA, Grey HM: HLA DR4w4-binding motifs illustrate the biochemical basis of degeneracy and specificity in peptide-DR interactions. J Immunol 1993;151:3163–3170.

4 Allen PM, Matsueda GR, Evans RJ, Dunbar JB, Marshall GR, Unanue ER: Identification of the T cell and Ia contact residues of a T cell antigenic epitope. Nature 1987;327:713–715.

5 Sinigaglia F, Hammer J: Defining rules for the peptide-MHC class II interaction. Curr Opin Immunol 1994;6:52–56.

6 Rötzschke O, Falk K: Naturally-occurring peptide antigens derived from the MHC class I re-stricted pathway. Immunol Today 1991;12:447–455.

7 von Boehmer H: Developmental biology of T cells in T cell-receptor transgenic mice. Annu Rev Immunol 1990;8:531–556.

8 Miller JFAP, Moroham G: Peripheral T cell tolerance. Annu Rev Immunol 1992;10:51–69.

9 Twari JL, Terasaki PI: HLA and Disease Association. New York, Springer, 1985.

10 Adorini L, Muller S, Cardinaux F, Lehmann PV, Falcioni F, Nagy ZA: In vivo competition between self peptides and foreign antigens in T-cell activation. Nature 1988;334:623–625.

11 Adorini L, Appella E, Doria G, Nagy ZA: Mechanisms influencing the immunodominance of T cell determinants. J Exp Med 1988;168:2091–2104.

12 Muller S, Adorini L, Juretic A, Nagy AZ: Selective in vivo inhibition of T cell activation by class II MHC-binding peptides administered in soluble form. J Immunol 1990;145:4006–4011.

13 Lamont AG, Powell MF, Colón SM, Mile C, Grey HM, Sette A: The use of peptide analogues with improved stability and MHC binding capacity to inhibit antigen presentation in vitro and in vivo. J Immunol 1990;144:2493–2498.

14 Rock KL, Benacerraf B: Inhibition of antigen-specific T lymphocyte activation by structurally related Ir gene-controlled polymers: evidence of specific competition for accessory cell antigen presentation. J Exp Med 1983;157:1618–1634.

15 Werdelin O: Chemically related antigens compete for presentation by accessory cells to T cells. J Immunol 1982;129:1883–1891.

16 Wraith DC, Smilek DE, Mitchell DJ, Steinman L, McDevitt HO: Antigen recognition in auto-immune encephalomyelitis and the potential for peptide-mediated immunotherapy. Cell 1989;59:247–255.

17 Lamont AG, Sette A, Fujinami R, Colón SM, Miles C, Grey HM: Inhibition of experimental autoimmune encephalomyelitis induction in SJL/J mice by using a peptide with high affinity for IAˢ molecules. J Immunol 1990;145:1687–1693.

18 Sakai K, Zamvil SS, Mitchell DJ, Hodgkinson S, Rothbard JB, Steinman L: Prevention of ex-perimental encephalomyelitis with peptides that block interaction of T cells with major histo-compatibility complex proteins. Proc Natl Acad Sci USA 1989;86:9470–9474.

19 O'Sullivan D, Arrhenius T, Sidney J, del Guercio MF, Albertson M, Wall M, Oseroff C, South-wood S, Colón SM, Gaeta FCA, Sette A: On the interaction of promiscuous antigenic peptides with different DR alleles. Identification of common structural motifs. J Immunol 1991;147:2663–2669.

20 Jardetzky T, Gorga JC, Busch R, Rothbard J, Strominger JL, Wiley DC: Peptide binding to HLA DR1, a peptide with most residues substituted to alanines retains MHC binding. EMBO J 1990;9:1797–1803.

21 Ishioka GY, Adorini L, Guery JC, Gaeta FCA, LaFond R, Alexander J, Powell MF, Sette A, Grey HM: Failure to demonstrate long-lived MHC saturation both in vitro and in vivo. J Immu-nol 1994;152:4310–4319.

22 Harding CV, Roof RW, Unanue ER: Turnover of Ia-peptide complexes is facilitated in viable antigen-presenting cells: Biosynthetic turnover of Ia vs. peptide exchange. Proc Natl Acad Sci USA 1989;86:4230–4234.

23 Adorini L, Appella E, Doria G, Cardinaux F, Nagy ZA: Competition for antigen presentation in living cells involves exchange of peptides bound by class II molecules. Nature 1989;342:800–803.

24 Pernis B: Internalization of lymphocyte membrane components. Immunol Today 1985;6:45–51.

25 Banga AK, Chien YW: Systematic delivery of therapeutic peptides and proteins. Int J Pharm 1988;48:15–19.

26 Langer R: New methods of drug delivery. Science 1990;249:1527–1533.

27   Hurtenbach U, Lier E, Adorini L, Nagy ZA: Prevention of autoimmune diabetes in non-obese
      diabetic mice by treatment with a class II major histocompatibility complex-blocking peptide.
      J Exp Med 1993;177:1499–1504.
28   De Magistris MT, Alexander J, Coggeshall M, Altman A, Gaeta FCA, Grey HM, Sette A:
      Antigen analog/major histocompatibility complexes act as antagonists of the T cell receptor.
      Cell 1992;68:625–634.
29   Ostrov D, Krieger J, Sidney J, Sette A, Concannon P: T cell receptor antagonism mediated by
      interaction between T cell receptor junctional residues and peptide antigen analogues. J Immu-
      nol 1993;150:4277–4283.
30   Alexander J, Snoke K, Ruppert J, Sidney J, Wall M, Southwood S, Oseroff C, Arrhenius T,
      Gaeta FCA, Colón SM, Grey HM, Sette A: Functional consequences of engagement of the T
      cell receptor by low affinity ligands. J Immunol 1993;150:1–7.
31   Krieger JI, Karr RW, Grey HM, Yu WY, O'Sullivan D, Batovsky L, Zheng ZL, Colón SM,
      Gaeta FCA, Sidney J, Albertson M, del Guercio MF, Chesnut RW, Sette A: Single amino acid
      changes in DR and antigen define residues critical for peptide-MHC binding and T cell recog-
      nition. J Immunol 1991;146:2331–2340.
32   Ruppert J, Sidney J, Celis E, Kubo RT, Grey HM, Sette A: Prominent role of secondary anchor
      residues in peptide binding to HLA-A2.1 molecules. Cell 1993;74:929–937.
33   Teale JM, Klinman NR: Tolerance as an active process. Nature 1980;288:385–387.
34   Klinman NR: The mechanism of antigenic stimulation of primary and secondary clonal precur-
      sor cells. J Exp Med 1972;136:241–260.
35   Jenkins MK, Schwartz RH: Antigen presentation by chemically modified splenocytes induces
      antigen-specific T cell unresponsiveness in vitro and in vivo. J Exp Med 1987;165:302–319.
36   Quill H, Schwartz RH: Stimulation of normal inducer T cell clones with antigen presented by
      purified Ia molecules in planar lipid membranes: Specific induction of a long-lived state of pro-
      liferative nonresponsiveness. J Immunol 1987;138:3704–3712.
37   Jenkins MK, Ashwell JD, Schwartz RH: Allogeneic non-T spleen cells restore the responsive-
      ness of normal T cell clones stimulated with antigen and chemically modified antigen-present-
      ing cells. J Immunol 1988;140:3324–3330.
38   Mueller DL, Jenkins MK, Schwartz RH: An accessory cell-derived costimulatory signal acts in-
      dependently of protein kinase C activation to allow T cell proliferation and prevent the induc-
      tion of unresponsiveness. J Immunol 1989;142:2617–2628.
39   Schwartz RH: A cell culture model for T lymphocyte clonal anergy. Science 1990;248:
      1349–1356.
40   Desilva DR, Urdahl KB, Jenkins MK: Clonal anergy is induced in vitro by T cell receptor occu-
      pancy in the absence of proliferation. J Immunol 1991;147:3261–3267.
41   Lamb JR, Skidmore BJ, Green N, Chiller JM, Feldmann M: Induction of tolerance in influenza
      virus-immune T lymphocyte clones with synthetic peptides of influenza hemagglutinin. J Exp
      Med 1983;157:1434–1447.
42   Sloan-Lancaster J, Evavold BD, Allen PM: Induction of T-cell anergy by altered T-cell-receptor
      ligand on live antigen-presenting cells. Nature 1993;363:156–159.
43   Ruppert J, Alexander J, Snoke K, Coggeshall M, Herbert E, McKenzie D, Grey HM, Sette A:
      Effect of T-cell receptor antagonism on interaction between T cells and antigen-presenting cells
      and on T-cell signaling events. Proc Natl Acad Sci USA 1993;90:2671–2675.
44   Evavold D, Allen PM: Separation of IL-4 production from Th cell proliferation by an altered
      T cell receptor ligand. Science 1991;252:1308–1310.
45   O'Rourke AM, Mescher MF: Cytotoxic T-lymphocyte activation involves a cascade of signal-
      ling and adhesion events. Nature 1992;358:253–255.
46   Brown JH, Jardetzky TS, Gorga JC, Stern LJ, Urban RG, Strominger JL, Wiley DC: Three-di-
      mensional structure of the human class II histocompatibility antigen HLA-DR1. Nature
      1993;364:33–39.
47   Lane DP: p53, guardian of the genome. Nature 1992;358:15–16.
48   El-Deiry WS, Kern SE, Pietenpol JA, Kinzler KW, Vogelstein B: Definition of a consensus
      binding site for p53. Nat Genet 1992;1:45–50.

49 Jameson SC, Carbone FR, Bevan MJ: Clone-specific T cell receptor antagonists of major histo-compatibility complex class I-restricted cytotoxic T cells. J Exp Med 1993;177:1541–1550.

50 Racioppi L, Ronchese F, Matis LA, Germain RN: Peptide-major histocompatibility complex class II complexes with mixed agonist/antagonist properties provide evidence for ligand-related differences in T cell receptor-dependent signaling. J Exp Med 1993;177:1047–1060.

51 Kuchroo VK, Sobel RA, Laning JC, Martin CA, Greenfield E, Dorf ME, Lees MB: Experimental allergic encephalomyelitis mediated by cloned T cells specific for a synthetic peptide of myelin proteolipid protein: Fine specificity and T cell receptor Vβ usage. J Immunol 1992;148:3776–3782.

52 Esch T, Clark L, Zhang XM, Goldman S, Heber-Katz E: Observations, legends, and conjectures concerning restricted T-cell receptor usage and autoimmune disease. Crit Rev Immunol 1992;11:249–264.

53 Urban JL, Kumar V, Kono DH, Gomez C, Horvath SJ, Clayton J, Ando DG, Sercarz EE, Hood L: Restricted use of T cell receptor V genes in murine autoimmune encephalomyelitis raises possibilities for antibody therapy. Cell 1988;54:577–592.

54 Sakai K, Sinha AA, Mitchell DJ, Zamvil SS, Rothbard JB, McDevitt HO, Steinman L: Involvement of distinct murine T-cell receptors in the autoimmune encephalitogenic response to nested epitopes of myelin basic protein. Proc Natl Acad Sci USA 1988;85:8608–8612.

55 Burns FR, Li XB, Shen N, Offner H, Chou YK, Vandenbark AA, Heber-Katz E: Both rat and mouse T cell receptors specific for myelin basic protein use similar Vα and β chain genes even though the major histocompatibility complex and encephalitogenic determinants being recognized are different. J Exp Med 1989;168:27–39.

56 Wucherpfennig KW, Ota K, Endo N, Seidman JG, Rosenzweig A, Weiner HL, Hafler DA: Shared human T cell receptor Vβ usage to immunodominant regions of myelin basic protein. Science 1990;248:1016–1019.

57 Ben-Nun A, Liblau RS, Cohen L, Lehmann D, Tournier-Lasserve E, Rosenzweig A, Zhang JW, Raus JCM, Bach MA: Restricted T-cell receptor Vβ gene usage by myelin basic protein-specific T-cell clones in multiple sclerosis: Predominant genes vary in individuals. Proc Natl Acad Sci USA 1991;88:2466–2470.

58 Howell MD, Diveley JP, Lundeen KA, Esty A, Winters ST, Carlo DJ, Brostoff SW: Limited T-cell receptor β-chain heterogeneity among interleukin-2 receptor-positive synovial T cells suggests a role for superantigen in rheumatoid arthritis. Proc Natl Acad Sci USA 1991;88:10921–10925.

59 Paliard X, West SG, Lafferty JA, Clements JR, Kappler JW, Marrack P, Kotzin BL: Evidence for the effects of a superantigen in rheumatoid arthritis. Science 1991;253:325–329.

60 Snoke K, Alexander J, Franco A, Smith L, Brawley JV, Concannon P, Grey HM, Sette A, Wentworth P: The inhibition of different T cell lines specific for the same antigen with TCR antagonist peptides. J Immunol 1993;151:6815–6821.

61 Smilek DE, Wraith DC, Hodgkinson S, Dwivedy S, Steinman L, McDevitt HO: A single amino acid change in a myelin basic protein peptide confers the capacity to prevent rather than induce experimental autoimmune encephalomyelitis. Proc Natl Acad Sci USA 1991;88:9633–9637.

62 Jorgensen JL, Reay PA, Ehrich EW, Davis MM: Molecular components of T-cell recognition. Annu Rev Immunol 1992;10:835–873.

63 Ehrich EW, Devaux B, Rock EP, Jorgensen JL, Davis MM, Chien YH: T cell receptor interaction with peptide/major histocompatibility complex (MHC) and superantigen/MHC ligands is dominated by antigen. J Exp Med 1993;178:713–722.

64 Sette A, Sidney J, Albertson M, Miles C, Colón SM, Pedrazzini T, Lamont AG, Grey HM: A novel approach to the generation of high affinity class II binding peptides. J Immunol 1990;145:1809–1813.

65 Wauben MHM, Boog CJP, van der Zee R, Joosten I, Schlief A, van Eden W: Disease inhibition by major histocompatibility complex binding peptide analogues of disease-associated epitopes: More than blocking alone. J Exp Med 1992;176:667–677.

66   Oki A, Sercarz E: T cell tolerance studied at the level of antigenic determinants. I. Latent reactivity to lysozyme peptides that lack suppressogenic epitopes can be revealed in lysozyme-tolerant mice. J Exp Med 1985;161;897–911.
67   Ria F, Chan BMC, Scherer MT, Smith JA, Gefter ML: Immunological activity of covalently linked T-cell epitopes. Nature 1990:343;381–383.
68   Gaur A, Wiers B, Liu A, Rothbard J, Fathman CG: Amelioration of autoimmune encephalomyelitis by myelin basic protein synthetic peptide-induced anergy. Science 1992;258:1491–1494.
69   Valli A, Sette A, Kappos L, Oseroff C, Sidney J, Miescher G, Hochberger M, Albert ED, Adorini L: Binding of myelin basic protein peptides to human histocompatibility leukocyte antigen class II molecules and their recognition by T cell from multiple sclerosis patients. J Clin Invest 1993;91:616–628.
70   Ota K, Matsui M, Milford EL, Mackin GA, Weiner HL, Hafler DA: T-cell recognition of an immunodominant myelin basic protein epitope in multiple sclerosis. Nature 1990;346:183–187.
71   Pette M, Fujita K, Wilkinson D, Altmann DM, Trowsdale J, Giegerich G, Hinkkanen A, Epplen JT, Kappos L, Wekerle H: Myelin autoreactivity in multiple sclerosis: Recognition of myelin basic protein in the context of HLA-DR2 products by T lymphocytes of multiple-sclerosis patients and healthy donors. Proc Natl Acad Sci USA 1990;87;7968–7972.
72   Page DM, Alexander J, Snoke K, Appella E, Sette A, Hedrick SM, Grey HM: Negative selection of CD4+ CD8+ thymocytes by T-cell receptor peptide antagonists. Proc Natl Acad Sci USA 1994;91:4057–4061.
73   Pircher H, Rohrer UH, Moskophidis D, Zinkernagel RM, Hengartner H: Lower receptor avidity required for thymic clonal deletion than for effector T-cell function. Nature 1991;351:482–485.
74   Vasquez NJ, Kaye J, Hedrick SM: In vivo and in vitro clonal deletion of double-positive thymocytes. J Exp Med 1992;175:1307–1316.
75   Ashton-Rickard PG, Bandeira A, Delaney JR, Van Kaer L, Pircher HP, Zinkernagel RM, Tonegawa S: Evidence for a differential avidity model of T cell selection in the thymus. Cell 1994;76:651–663.
76   Allen PM: Peptides in positive and negative selection: a delicate balance. Cell 1994;76:593–596.
77   Hogquist KA, Jameson SC, Heath WR, Howard JL, Bevan MJ, Carbone FR: T cell receptor antagonist peptides induce positive selection. Cell 1994;76:17–27.
78   Spain LM, Jorgensen JL, Davis MM, Berg LJ: A peptide antigen antagonist prevents the differentiation of T cell receptor transgenic thymocytes. J Immunol 1994;152:1709–1717.

Alessandro Sette, PhD, Cytel, 3525 John Hopkins Court, San Diego, CA 92121 (USA)

Adorini L (ed): Selective Immunosuppression: Basic Concepts and Clinical Applications.
Chem Immunol. Basel, Karger, 1995, vol 60, pp 79–99

....

Modified T-Cell Receptor Ligands: Moving beyond a Strict Occupancy Model for T-Cell Activation by Antigen

Luigi Racioppi [a], *Ronald N. Germain* [b, 1]

[a] Laboratory of Immunology, Department of Cellular and Molecular Biology and Pathology, University of Naples 'Federico II', Naples, Italy;
[b] Lymphocyte Biology Section, Laboratory of Immunology, National Institute of Allergy and Infectious Disease, National Institute of Health, Bethesda, Md., USA

The Molecular Basis of T-Cell Recognition

The primary physiological role of major histocompatibility complex (MHC) molecule is the sampling of peptides derived from different intracellular pools, and the stable display of these acquired antigenic peptides in a multivalent form on the surface of the cell [1, 2]. Antigen-specific activation of CD4 and CD8 T lymphocytes depends on the subsequent binding of these MHC molecule-peptide complexes by the clonally distributed αβ-receptors on these lymphocytes, resulting in the initiation of second messenger generation and development of effector activity. Understanding the molecular details of TCR-ligand interaction and how TCR recognition is translated into intracellular signals mediating T-cell activation have thus been major goals of immunological research over the past several years.

Very substantial progress has been made in the area of MHC structure and peptide presentation. MHC molecules are highly polymorphic proteins,

[1] We wish to thank Prof. Serafino Zappacosta, Dr. Anna Maria Masci, Dr. Giuseppe Matarese and members of the Zappacosta laboratory for their help and useful suggestions during the preparation of this manuscript. Dr. Racioppi was supported in part by UICC, Geneva, ICRETT award no. 788.

and their structural variation has been shown to play a critical role in both the mature immune response [3] and the early development of T cells [4, 5]. Two major mechanisms by which this polymorphism affects the specificity of T-cell responses have been delineated. First, different MHC alleles have distinct affinities for individual peptides. Sequencing of peptides eluted from MHC molecules has revealed allele-specific motifs, corresponding to critical anchor residues that fit into specific pockets in the MHC peptide binding groove. The chemistry of these binding pockets is dictated largely by the side chains of residues at polymorphic positions in the protein [6]. Second, in addition to these quantitative effects on peptide binding, MHC polymorphic residues can impose allele-specific conformations on the peptide-MHC complex that can result in marked effects on T-cell response [7].

Advances have also occured in our knowledge about TCR recognition of peptide-MHC complexes. Recent studies on the crystal structure of MHC peptide complexes indicate that only three or four amino acid side chains of the immunogenic peptide are accessible to the TCR. An indirect strategy to identify these crucial residues necessary for T-cell receptor engagement and signaling has been the use of analogues of the immunogenic peptide carrying selected amino acid substitutions [8–12]. These analogues have been tested both for their ability to induce T-cell responses and to bind MHC molecules. This experimental approach has provided data suggesting a hierarchy in the importance of the epitopic (exposed) residues of an immunogenic peptide with respect to TCR signaling. The T-cell antigen receptor seems to focus most intensely on a 'primary' contact residue that is critical for stimulation of most of the T cells elicited by a given antigenic determinant [12].

A large number of sequence studies have shown that when a particular peptide-MHC complex is recognized, the CDR3 sequences of the TCR α- and/or β-chain are often conserved, and evidence for a direct involvement of CDR3 in contacting the 'primary residues' of the immunogenic peptide has been obtained [14–20]. One recent study suggests that TCR recognition of the MHC-peptide complexes may be 'peptide centered' in that CDR3-peptide interactions can apparently affect the ultimate configuration of the trimolecular complex [21]. This model suggests that distant TCR/MHC contacts can be affected by changes in CDR3/peptide contacts. This implies that the same TCR might interact with the same MHC in different ways by use of different 'angles of approach', and that there is a relative fluidity to the precise intramolecular interactions between a given TCR and MHC-peptide complex. The existence of allele-specific MHC-peptide conformations and the dominant role played by the peptide in the control of TCR affinity and/or the 'angles of approach' in TCR binding reveal an unexpected level of complexity in this molecular interaction.

T-Cell Signal Transduction and Costimulation

Signals essential to T-cell triggering are elicited through the ligand-engaged TCR as a consequence of its interaction with other membrane proteins having the capacity to generate intracellular second messengers [22]. Recent data indicate that the CD3 γ, δ, ε:ζ/η complex stably associated with the αβ TCR consists of a minimum of two signal transduction modules with distinct and independent capacities to initiate second messenger cascades [23, 24] that include primary events such as tyrosine phosphorylation [25] and secondary events such as phosphoinositol bis-phosphate [PIP_2] hydrolysis, elevation of [CA^{2+}]$_i$ [26, 27], production of cAMP [28] and activation of specific serine-threonine kinases [29]. The effects of these biochemical events on T-cell differentiation are heavily influenced by other receptor-ligand interactions generating signals differing qualitatively from those evoked by TCR-ligand interaction [30–34]. The participation of such costimulatory signals is most evident in the control of IL-2 production [35], the cytokine principally responsible for the clonal expansion that follows T-cell activation. Metabolically inactivated cells bearing recognizable ligands for TCRs on CD4+ cells are poor or ineffective stimulators of IL-2-dependent T-cell proliferation [36–38]. The component missing in such circumstances contributes to T-cell IL-2 production by a mechanism independent of augmenting TCR occupancy and TCR-regulated second-messenger generation [39]. The best candidates for providing the critical costimulatory signal include CD80 and B70 [32–34, 40, 41], two counter-receptors for CD28 and CTLA-4, as well as the heat-stable antigen that binds an uncharacterized ligand on the T cell [42, 43]. This costimulatory apparatus seems to be regulated in a very precise manner during the immune response [44].

Despite the complex biochemical events associated with TCR-dependent intracellular signaling, a relatively simple model for the relationship between peptide-MHC ligand and TCR in the initiation of T-cell triggering has until recently been the primary paradigm in the field [45, 46; but see 47]. Activation was generally believed to occur when an adequate number of TCRs were occupied, generating signal competent aggregates at the adhesion receptor-generated interface between the T cell and the presenting cell [48]. The extent of this ligand-induced TCR aggregation was assumed to depend solely on the number of available receptors on the T cells, the number of available peptide-MHC complexes on the APC, and the affinity of the TCR for the ligand. It was commonly assumed that when high levels of MHC-peptide complexes on the APC failed to induce measurable T-cell activation, this was due to a low affinity of the TCR for the ligand that prevented receptor occupancy from exceeding the threshold needed for aggregation and second-messenger generation within the T cell itself [49]. This affinity-based occupancy model predicted that in the presence of intact, metabolically active APC known to be capable of delivering costimulatory signals, peptide-MHC complexes should fall into only two categories: complete antagonists (cAG) that can induce full T-cell activation and nonagonists (nAG) that are ignored by the T cell because even at saturation of the surface MHC molecules of the APC with peptide antigen, the number of occupied TCR would be below the triggering threshold, due to a low TCR affinity for the peptide-MHC complex [46].

New Classes of T-Cell Receptor Ligands

Recently, several observations have cast into doubt the validity of the affinity-based model of TCR occupancy described above. Parallel analyses of different parameters modulated by the process of T-cell activation (cytokine secretion, surface molecule expression, T-cell proliferation, cytotoxic activity) has revealed phenomena that cannot be explained by assuming only two categories of TCR ligands such as agonists and nonagonists.

The first report of TCR ligands with unpredicted characteristics came from Evavold and Allen [50]. They studied IL-4 production, cell proliferation, and B-cell helper function of amurine Th2 clone specific for Hb(64–76)/E^k complexes, and found that certain Hb peptide analogs containing a single residue substitution failed to stimulate proliferative responses, while eliciting IL-4 cytokine production and B-cell help. This indicated that a ligand could stimulate T-cell second-messenger generation without evoking the full repertoire of effector responses of which the T cell was capable. Careful dose-response analyses showed that this could not be explained as a being due simply to a different quantitative requirement for elicitation of IL-4 production vs. induction of proliferation. Thus, these MHC-peptide complexes had neither the properties of cAG nor nAG, but rather defined a new category of TCR ligands that could best be called partial agonists (pAG). The same laboratory extended these observations to Th1 clones, finding that some cAG analogs induced IL-2-Rα upregulation, T-cell size enlargement, and cytotoxic activity, in the absence of lymphocyte proliferation [51]. Most recently, they reported that some TCR ligands display pAG activity seen as an ability to anergize a Th1 clone in the presence of fully competent live APC, without stimulating a measurable effector response [52]. In this system, the strategy used to generate pAG peptides was the substitution of amino acid residues involved in TCR binding. Analogs containing substitutions in 'secondary' residues were able to work as pAG, whereas substitutions in 'primary' contact residues seems to generate peptides with nAG properties.

These reports of pAG were followed by studies of DeMagistris et al. [53], who found that some nAG peptides that were singly substituted versions of cAG peptides had the ability to inhibit responses in the presence of the cAG. These analogs caused a dose-dependent inhibition of T-cell proliferation when added to APC that were already displaying immunogenic (cAG) peptide-MHC complexes. This class of analog did not delivery any identifiable signal to the T cells. From these results, the authors hypothesized that analog-MHC complexes were acting as TCR antagonists by binding to receptors without invoking signal transduction and competitively preventing occupancy of these receptors with cAG peptide-MHC complexes. This new class of TCR ligands was considered to comprise complete receptor antagonists (cANT). To define the structural basis for cANT function, a large panel of peptides was generated by systematic substitution of each cAG residue or by sequential reincorporation of TCR contact residues into a polyalanine backbone [54]. While no obvious pattern of amino acid substitution predicted the generation of antagonists, these investigators observed that the most potent cANT were peptides very similar to the original cAG peptide, and they suggested that having a lower affinity than cAG complexes for the TCR was the critical feature of cANT complexes.

A third class of TCR ligands with mixed agonist/antagonist properties was observed in our own studies of a murine alloantigen model [55]. In this study, an alternative strategy to the above peptide analog approach was used to manipulate the interaction of the MHC-peptide complexes with the TCR. TCR ligand analogs were obtained by site-direct mutagenesis of the murine I-E^k molecule in the HV3 region of β-chain, a region previously identified as involved in control of the quality and quantity of peptide binding and also TCR interaction [7]. The effects of these changes were studied using a Th1 clone specific for the cytochrome c(81–104)/E^k complex that was also alloreactive for the closely related E^s molecule. k→s substitutions of amino acids at positions β75 and β79 (E^{kss}) were sufficient to confer allostimulatory function on the mutated molecule. Incubation of cells expressing E^{kss} molecules with the immunogenic peptide c(81–104) surprisingly led to selective inhibition of IL-2 production without interference with alloantigenic stimulation of IL-3 production, IL-2Rα upregulation, or cell-size enlargement. The inhibitory effect was not observed with using unrelated peptides displaying comparable affinity for E^{kss} molecules, and occurred despite the continued presence of alloantigenic complexes on the APC. An analog of c(81104) substituted in position 99 (99Q) maintained its ability to bind E^k and E^{kss} molecules but lacked the capacity to act either as an agonist or antagonist. Interestingly, position 99 of the cytochrome c peptide has been shown to be a 'primary' TCR contact residue [14]. These data confirmed that c(81–104)/E^{kss} complexes were able to selectively modulate certain effector activities in a TCR-specific manner, and that the effect of the added peptide was not due to competition for MHC molecule occupancy.

There thus seemed to be three different classes of variant TCR ligands partial agonists [50] complete antagonists [53] and mixed partial agonists/partial antagonists [55]. The latter superficially differ from the partial agonists in that they selectively inhibited certain effector responses as opposed to simply failing to stimulate these responses. At the same time, these ligands seemed to differ from complete antagonists in that they selectively inhibited certain effector responses without affecting other ones, and acted without preventing all measurable T-cell signaling. More recent studies, however, suggest that all three types of ligand may operate in a related manner, and represent parts of a continuum from nearly complete agonist function to nearly complete antagonist function. Thus, the ligands studies by DeMagistris et al. fail to block some TCR-based signals induced by cAG such as those promoting cell adhesion, thus falling into the category of partial antagonists, whereas some of the ligands studied by Allen and colleagues show not only partial agonist, but also partial antagonist activity.

The phenomena of TCR antagonism has also been extended to CD8+ T lymphocytes [56]. Most of the antagonists identified in this system contained substitutions at TCR contact sites. Some clearly fell into the category of mixed partial agonists/antagonists, whereas others had predominantly antagonist function. Even in this case, however, some variation in effect on different activation parameters was observed, suggesting that as in the above examples, these pAG and cANT ligands represent polar extremes on a continuum. It is thus clear that modified receptor ligands (MRL) have widespread biological activity that affect both major subclasses of $\alpha\beta$ T lymphocytes.

A Model for TCR-Ligand Interactions

The molecular mechanism by which a modified receptor ligand can change the integrated response to TCR signaling is unclear. A simple view of the occupancy model of T-cell activation would argue that independent of affinity, any ligand able to engage the receptor should contribute in a positive sense to the development of intracellular signals. This is because the presence of ligand of any quality would always serve to increase the steady state number of occupied receptors and thus the number of aggregated TCR complexes. Furthermore, no matter what the affinity of the interaction with the TCR, ligands that do not reduce the available number of agonist complexes on the APC itself should not interfere with T-cell activation by the good ligand. This is because the poor ligand must at equilibrium itself occupy each TCR that is prevented from interacting with the good (agonist) ligand. Therefore, although the nature of the ligand occupying the TCR will change as more and more poor ligand is added, the total number of occupied receptors will not be altered and the same aggregation-dependent signal generation should occur.

Operationally, this model conceives of each TCR molecule as a functional unit that can assume only two discrete values. When the TCR is occupied by a ligand it is in an 'ON' position and can contribute to aggregates that lead to transduction of a positive signal. Alternatively, an unoccupied TCR is in an 'OFF' position, and has no positive or negative effect on the T cell. TCR-generated positive signals are integrated at an intracellular level and T-cell responses occur when a given level (threshold) of occupancy-generated signals has been achieved. No physical or kinetic limits are imagined to exist that can affect the integration of TCR signaling, so that the time during T cell/APC interaction at which a particular receptor is occupied has no qualitative influence on T-cell signaling or activation and the whole repertoire of the plasma membrane-expressed molecules are available to equally participate in signal generation. As this simple model is not able to explain the biological responses induced by MRL peptides, a more complex model must be formulated.

Consideration of possible physical limits to the process governing the intracellular integration of TCR-signaling is one of the first steps necessary to derive a more suitable model of T-cell activation. A stable T-cell/APC interaction requires recognition and binding of MHC-peptide complexes by the antigen T-cell receptors. The effective engagement of TCR by peptide-MHC molecule complexes appears to first require membrane-related intermolecular interactions and signaling events involving coreceptors and adhesion molecules (M signals), followed by intracellular or nuclear signaling (N signals) resulting in differential gene activation [57]. M signals result in the formation of synapses-like regions between the T cell and APC in the area of initial contact. These synapses-like zones can be imagined to function as microdomains where antigen presentation occurs. Thus, their extent and molecular composition would impose topographic limits to an integrative signaling process. Molecules that are not localized to or that fail to enter these active zones would not contribute to the activation process.

While effective M signaling seems to be a condition necessary to establish functional T/APC interactions, N signaling gives rise to the T-cell biological response. Well-characterized aspects of N signaling comprise the triggering of the second-messenger cascade [58] and the relocalization of several tyrosine and serine/threonine kinases in the proximity of ligand-engaged TCR molecules [59]. This network of intracellular signals presumably involves both positive and negative biochemical feedback loops, and results in a self-limited

T-cell activation process. The kinetics of reactions generating the negative mediators would impose time limits on the process that integrates TCR-derived signals for positive effector responses. TCR ligands with agonist activity would generate 'balanced' M and N signaling that permits the necessary positive signals to develop and propagate in the cell interior (to the nucleus) without interference from the negative regulatory limbs of the signaling system. This model of T-cell activation postulates a hierarchy of signals involved in the mechanism of lymphocyte activation, with cAG and pAG ligands distinguished from ANT ligands by the inability of the latter to either elicit M signals on their own or to interfere with such signals in the presence of cAG [60].

Recently, X-ray analysis of DR1 molecules revealed the existence of parallel pairs of DR1 dimers in three different crystallized molecules. Based on this observation, Brown et al. [61] proposed that the clustering of two or more TCR $\alpha\beta$ dimers would be required to produce a functional signaling complex and suggested that stable formation of MHC superdimers with two identical peptides was dependent on interaction with suitable TCRs. If a TCR in contact with an MHC-bearing cell has negligible affinity for any of the peptide-MHC complexes on the cell surface, these various surface proteins will exist as separate entities and the signal generated by the few, transient binary complexes that form would not be sufficient to activate the T cell. However, in the presence of a peptide-MHC combination recognized with suitable affinity, monomeric TCRs bound to these cognate ligands would begin to accumulate at the synapses-like region of contact between T cell and APC. As the concentration of the monovalent TCR/peptide-MHC ligand complexes increases in this interface, they would begin to contact one other, allowing bonds to form along the potential superdimer interface and between TCR units. The lateral binding of TCR to TCR and MHC to MHC would generate a stabilized complex by converting low affinity interactions into a high avidity association by a 'chelating' effect. The stable dimerization of MHC molecules could also allow them to generate a signal within the antigen presenting cell. The 'chelating' effect would also involve CD4/CD8 molecules that undergo co-aggregation with TCR/MHC complexes.

This model suggests that TCR need to dimerize to be functionally active. However, it is not possible to exclude that the 'chelating' process involves a larger number of TCR molecules and results in larger functional units (TCR microaggregates) [62]. The density of MHC-peptide complexes in the synapse-like zones and their affinity for the TCR would determine the statistical probability of stabilizing functionally active TCRs microaggregates. We propose that irrespective of the absolute number of TCR occupied, only certain stable geometric configurations of TCR microaggregates would be able to transduce 'balanced' signals that evoke a complete but homeostatically regulated T-cell

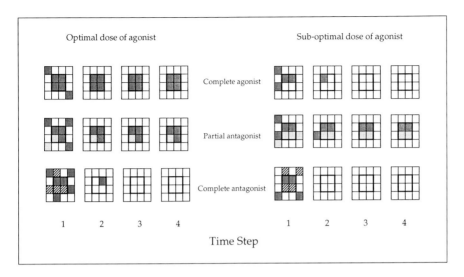

Fig. 1. Hypothetical model of TCR signaling. The synapse-like zone is represented as a dynamic system, the evolution of which is completely described by local interactions, and is discrete in both space and time. This defines a cellular automaton of sixteen sites. The rules governing this cellular automaton (CA) are as follows: (1) Each CA consists of a discrete lattice of sites (TCR molecules). (2) CA evolve in discrete time steps. (3) Each site takes on a finite set of possible values (unoccupied, agonist, partial antagonist, antagonist) that are represented in the scheme respectively as open, dark gray, light gray, and hatched squares. (4) The value of each site evolves according to the same 'deterministic' rules. (5) The rules for the evolution of a site depend only on the local neighborhood of sites around it. In this model, a gray site remains gray in the next step time only if at least three neighbors are gray sites. Alternatively, if these conditions are not respected, a gray site becomes an unoccupied (open) site in the next step time.

The four central sites included in the bold boards represent the basic functional unit of the transducing machinery. The scheme depicts the effects of MRL acting in the presence of optimal or suboptimal concentrations of an agonist ligand. In the presence of an optimal agonist concentration, the major effect of partial antagonists is to prevent stabilization of 'balanced' TCR microaggregate geometries. In the presence of suboptimal doses of agonist, they permit the stabilization of 'unbalanced' microaggregates. Complete antagonists prevent the stabilization of any effective microaggregate geometry.

response. Alternatively, 'unbalanced' geometries would generate a spectrum of signaling ranging between the selective transduction of M signals alone, to partial, or to full effector activation of T lymphocytes. This model can be described statistically as a 'cellular automaton', that is, a dynamic system, the evolution of which is completely described by local interactions, and whose function is discrete in space and time (fig. 1).

According to this hypothesis, MRLs should act by modifying the stability and/or the symmetry of TCRs microaggregates. This is consistent with quantitative data that imply that antagonist function cannot reflect simple competition for occupancy of individual unoccupied TCR, but rather, must reflect the interference by TCR-antagonist pairs with the effective signaling function of TCR-cAG pairs, i.e. that the competition is between occupied receptors for effective formation of the signaling microaggregates. This is most easily appreciated from the fact that only a small molar excess of antagonist need be present for inhibition of cAG stimulation at minimal levels of T-cell activation [53, 56]. Because it appears that under such conditions, TCR cAG occupancy is typically in the range of a few hundred receptors out of a total pool of >2–5×10^4, antagonist complexes only a few-fold higher than cAG complexes should only engage perhaps a thousand or so TCR. This is clearly too low a number to decrease cAG stimulation by direct elimination of available *free* TCR binding sites. However, this number of ANT ligands could generate *occupied* TCR in severalfold molar excess over the number occupied with cAG ligand, so that any forming TCR microaggregate would include an ANT-occupied TCR that could disrupt stabilization of an effective signaling complex.

Antagonists could achieve such disruption of signal-competent microaggregates by affecting the on-off kinetics of the reaction between TCR and MHC-peptide complexes (perhaps by preventing stable dimerization), and in this way reduce stable trimolecular complexes available to form microaggregates. Alternatively, antagonists could impose architectural motifs on the forming microaggregates that were unsuitable for 'balanced' TCR signaling that depends on the proper spatial organization for signal coordination. Partial agonist ligands would trigger M signals and stabilize aggregate conformations able to activate only certain signaling subroutines, leading to a limited range of effector responses that could include anergy. For certain cellular responses, cAG/partial ANT mixtures on the APC would fail to generate substrates involved in the positive signaling while maintaining the capacity to trigger negative feedback pathways (fig. 2, 3). A prediction of this model is that a partial antagonist could act as a TCR agonist for certain cellular responses that are not affected by the negative mediators and at the same time be able to induce dominant negative signaling for other susceptible cellular responses (fig. 2, 3).

Several experimental observations appear to be in accord with these predictions. For example, in studies of a Th1 clone specific for the cytochrome c(81–104)/E^k complex that was also alloreactive for the closely related E^s molecule, addition of the original agonist peptide cytochrome c(81104) to L-cells expressing E^{kss} allostimulatory molecules selectively inhibited IL-2 production by this clone without preventing IL-3 production, IL-2Rα upregulation, or cell-size enlargement. A T-cell hybridoma derived by fusion of a

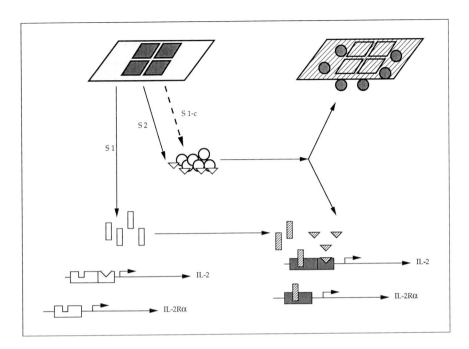

Fig. 2. Transduction of 'balanced' signals. TCR engagement with complete agonist ligands leads to stabilization of microaggregates transducing a full repertoire of activating
signals. It is proposed that TCR microaggregates induce at least two qualitatively distinct
signals, S1 and S1-c. S2 is generated by costimulatory and/or accessory molecules. S1 signaling acts on costimulatory unrelated functions, while S2 and S1-c jointly act on costimu-
lation-dependent T-cell responses (e.g. IL-2 production). S2 and S1-c display qualitatively
distinct biochemical effects. In this scheme, S1-c signaling induces topological relocalization of biochemical mediators (i.e. nuclear factors and enzymes) controlling the activation
of costimulation-related cellular responses and triggering of negative feedback loops. S2
signaling itself activates the same pattern of biochemical mediators irrespective of their
cellular localization. To obtain a full T-cell response, S1, S1-c, and S2 signaling need to be
synchronized in time and space. The dotted large tilted square represents the T-cell side of
a synapse-like zone generated in the T cell/APC contact area. The small squares represent
an agonist induced-TCR microaggregate in the early phases of T-cell activaton (gray) and
after triggering of negative feedback loops (hatched). Open circles, triangles and rectangles represent biochemical mediators of costimulation-related (circles and triangles) and
-unrelated (rectangles) functions. Closed circles and hatched triangles and rectangles
represent the activated forms of these biochemical mediators.

<hr />

Modified T-Cell Receptor Ligands

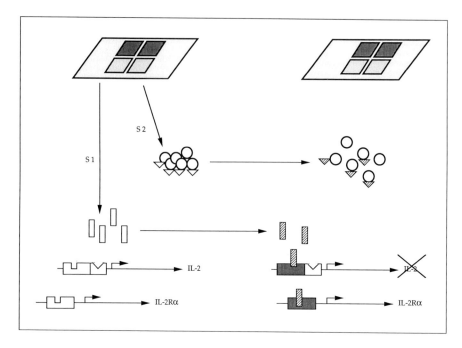

Fig. 3. Transduction of 'unbalanced' signals. In accord with the model proposed in the text, a partial antagonist would stabilize 'unbalanced' TCR microaggregates that fail to transduce S1-c signals. The biochemical mediators responsible for the costimulation related T-cell responses would be activated by S2 signals but they would fail to be complemented by S1-c-signal-dependent changes. This mechanism would lead to selective inhibition of costimulation-related T-cell functions, while permitting activation of S1-dependent, costimulation-independent functions.

TCR-negative T lymphoma cell with this same Th1 clone and possessing the same peptide and alloantigen response was also studied. With the hybridoma as the responding cell instead of the corresponding normal clone, it was found that at high concentrations the $c(81–104)/E^{kss}$ complex could clearly elicit a partial effector response as evidenced by increased IL-3 production. These data reveal a class of MRL displaying agonist activity for certain cellular responses and antagonist behavior for other T-cell functions, with the nature of the effect also being dependent on the cellular environment in which the signal transduction and effector responses were occurring.

In considering this model, it is important to discuss the question of whether only quantitative effects on overall signaling levels or qualitative effects on the nature of the signals generated are responsible for the selective stimulation of some responses by pAG or the selective inhibition of responses

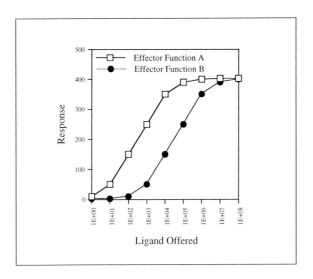

Fig. 4. Hypothetical dose-response curves for two different T-cell activities. Activity A requires 100-fold less cAG to be elicited than activity B. At dose X, the T cell will produce activity A but not B, whereas at dose cAG > than X, both activities will be induced. If a MRL is tested and shows no stimulation of activity A in the face of production of activity B, or if such a ligand produces A without B over a dose range greater than 100-fold, then the cell must be receiving qualitatively different signals in comparison to cAG, rather than just receiving less signal. Similarly, if a MRL inhibits cAG stimulation of A more potently than activity B, this ANT function cannot arise from simple competitive occupancy of TCRs and a parallel loss of all signals generated by cAG engagement.

by pANT. For example, could pANT act merely to decrease the overall level of cAG-induced signal generation, with the apparent selective effects of pANT resulting from different thresholds for the various cellular responses being measured (fig. 4). If this were the case, then the inhibition of cellular activity by pAG should follow a defined pattern, in which those responses requiring higher levels of cAG, and, hence, more signaling, should be the first to be inhibited as pANT complexes are formed. The converse would be true for pAG, with only the more sensitive response being initiated due to the less effective signal generation by such ligands. These scenarios do not easily fit with two experimental results. In the studies of Evavold and Allen, the Hb specific T cells show IL-4 and proliferation responses to cAG that when expressed as '% maximum response' follow similar dose-response curves; yet the variant Hb peptide (pAG) fails to stimulate proliferation at peptide concentrations several orders of magnitude greater than necessary for IL-4 secretion. In the case of the alloreactive clone we have studied, 100-fold more cytochrome c peptide is required with wild-type class II molecules to stimulate half-maximal IL-3 responses than IL-2 responses, but the pANT combination selectively inhibits IL-2, not IL-3. Thus, these data argue that simple competition for receptor occupancy, with a parallel decrease in all signaling parameters, is not an adequate explanation of how these MRL alter T-cell responses. Rather, there appear to be changes in the quality of the signals transduced, as predicted from the model presented above.

Biological Effects of Modified TCR Ligands

Although the molecular details by which MRL affect T-cell responses remain poorly understood, the biological phenomena mediated by these classes of TCR ligands seem to be extremely relevant to the control of both physiological and pathological immune responses. Certain classes of MRL are able to differentially affect the costimulatory-dependent and -independent effector responses of a T cell. In at least some cases, partial agonists seem to preferentially stimulate costimulatory-independent responses such a cytotoxic target cell lysis, while partial antagonists selectively inhibit the co-stimulatorydependent ones such as IL-2 secretion.

This linkage of partial antagonist function with costimulation-related processes raises the intriguing possibility that such TCR ligands may play a significant role in peripheral and central T-cell tolerance. Engagement of TCR in the absence of adequate costimulation causes Th1 cells to became unresponsive to subsequent antigen stimulation ('anergy') [35]. Recently, Sloan-Lancaster et al. [52] have shown that a partial agonist unable to stimulate clonal proliferation, the production of cytokines, or the generation of inositol phosphate metabolites can nonetheless induce cloned T cells to become profoundly unresponsive to subsequent stimulation with a cAG peptide. Induction of this anergic state occurred even when the APC were live lipopolysaccharide-treated spleen cell blasts, and was prevented by addition of cyclosporine A [52]. The identification of MRL able to induce the anergic state in T lymphocytes suggests that at least in part, T-cell peripheral tolerance could be maintained by self peptides displaying partial antagonist activity. Perhaps among the set of nested peptides presented by class II MHC molecules, some ligands will serve as ANT for the T cells seeing other overlapping peptides as cAG.

The T-cell repertoire is shaped by the processes of negative and positive selection that occur in the thymus. Both processes have been thought to involve the engagement of the TCR with MHC-peptide complexes. Positive selection results in survival of the cell and eventual differentiation into a functional lymphocyte. Negative selection results in activation-induced apoptotic death. Recent evidence suggest that self peptides are profoundly involved in the process of positive selection [63, 64]. Hogquist et al. [64] using organ cultures of fetal thymic lobes from TCR transgenic, β2M (–/–) mice have shown that peptides with antagonist behavior with respect to mature T-cell activation were the most effective in postively selecting CD8+ thymocytes [64]. This may be the result of the type of 'unbalanced' (partial agonist) signaling described above, such that differentiation but not apoptosis is elicited over a range of ligand concentrations. In contrast, the same levels of cAG complexes might co-induce both differentiation and death pathways, with the net effect being a failure to successfully mature. Based on the data of Ashton-Rickardt et al. [63] rather than relying on the spread between these parameters characteristic of partial agonist function, it also appears that the threshold differences illustrated in figure 4 for cAG could also achieve the same end.

When triggered by antigen, naive T cells produce IL-2 only, which acts as an autocrine growth factor, but after clonal expansion they become competent to produce a variety of lymphokines. Under the influence of the local accessory cell/cytokine microenvironment, activated T cells may differentiate to Th1 or Th2 cells [65]. Th1 cells produce IL-2, IFN-γ and other soluble factors inducing inflammatory responses. Th2 cells produce IL-4, IL-5 and IL-10 and help in the production of nonopsonizing immunoglobulins G, A and E. Many immune responses polarize toward either Th1 or Th2, causing protection or immunopathology in the face of certain stimuli. Polarization occurs because Th1 and Th2 cells make lymphokines that suppress the development and the effector functions of cells of the reciprocal type [66]. In several infections, such as leprosy, tuberculosis and AIDS, the protective Th1 response may be inhibited by the development of Th2 cells [67, 68]. However, in other conditions Th2 response and IgE play a protective role, whereas inflammatory response may cause immunopathology [69].

MRL could control Th1/Th2 choice by permitting or inducing production of certain type of lymphokines but not others. Furthermore, tissue-specific self peptides displaying partial antagonist or partial agonist activities could be involved in the control of regional Th1/Th2 immune-response phenotypes. Similarly, pathogens could evolve mechanisms that would allow them to escape immune detection if a high rate of genetic mutation led to the production of partial agonists or antagonists for the original immunodominant epitope.

Partial antagonist ligands can be generated experimentally not only by varying the structure of the MHC-associated peptide, but also by manipulating MHC molecules themselves through the introduction of point mutations in the binding domain [55]. This suggests that in physiological conditions, MHC molecules displaying high levels of structural homology might bind the same peptide and form complexes displaying antagonist or partial antagonist activity for each other. This phenomenon of simultaneous formation of both agonist and antagonist ligands on individual APC could explain how expression of certain MHC alleles decreases development of specific autoimmune diseases in the presence of a known risk-associated allele. In this case, the protective allele would be imagined to generate antagonist ligands upon self-peptide binding that depress the stimulatory effect of cAG complexes involving the disease-inducing allele. A variation on this theme implies that self peptide complexes on APC may in general contain sufficient numbers of antagonists for TCR reactive with foreign cAG complexes to blunt the initiation of immune responses to these foreign antigens.

Potential Therapeutic Applications of Modified TCR Ligands

The discovery of TCR ligands with unexpected biological properties provides new possibilities for the manipulation of immune responses. In vaccine development, partial agonist/antagonists might be designed to selectively guide the immune response along certain pathways, avoiding immune responses that can have pathological rather than beneficial effects, and enhancing stimulation of protective pathways. The ability of partial or complete antagonists to dominantly interfere with T-cell-effector function also suggests new approaches to autoimmune disease treatment. A problem with the common broadly active immunosuppressive agents such as steroids, cyclosporine A/FK 506, cyclophosphamide, or methotrexate is that they leave the patient more prone to infections and for the development of malignancies. They are also associated with a high level of undesirable side effects, such as renal and liver damage.

A more specific immunosuppressive strategy involves use of peptides that can physically block the MHC molecule binding sites and prevent presentation of the peptides involved in the disease [70]. This method requires massive amounts of material to achieve the necessary quantitative blocking effect, may be accompanied by induction of undesirable strong immune response to the blocking agent itself, and calls for continuous tretament because the effect on the T cells is not prolonged. Moreover, there is no evidence that peptides that block the MHC binding site can affect presentation of self antigens that are pre-associated with MHC molecules, which would be necessary for treatment when active disease is already present [71].

Complete antagonists avoid several of these problems as the antagonist are immunologically specific and thus affect only a small subset of T cells relevant to disease. This reduces the chance of adverse systemic or organ-specific side effects, and lessens the amount of material necessary for administration as compared to the MHC molecule blocking strategy. However, complete antagonists do not generate intracellular signals, which means that any effects of administration would be transient.

Partial agonists and partial antagonists should be better candidates than complete antagonists for effective induction of selective immunosuppression. This is because these classes of TCR ligands could potentially drive autoimmune T cells into a long-term unresponsive state [52]. Therefore, partial agonists/antagonists may potentially be able to block ongoing autoimmune T-cell effector activity, as might a complete MHC blocking peptide, and administration of such a TCR ligand may also lead to a lasting decrease in autoimmune disease due to anergy induction among the involved T cells. In this fashion, ligand administration could diminish or slow the progression of the autoimmune disease process. This would be accomplished with few or no side effects, due to extreme specificity of the drug for only the disease-causing T cells. But this same high degree of specificity also poses a problem for this approach, namely whether the entire cohort of TCR involved in the response to even simple, and more importantly, complex antigens would be susceptible to one or only a few such MRL. Preliminary data suggest that a cocktail of a few compounds might provide adequate coverage for the T cells responding to a given peptide cAG; the question will be whether in an active disease state, only one or a few such cAG are relevant. Alternatively, it might be possible for pAG to generate cells that when faced with the natural self antigen, secrete lymphokines that block the pathologic inflammatory response in the surrounding tissue; this mode of action would avoid the need to antagonize the stimulation of all the autoreactive T cells.

Conclusion

The finding of new classes of TCR ligands and the description of the biological phenomena mediated by these categories of molecules reveals an unpredicted level of complexity in the elicitation of T-cell immune responses. The simple affinity-based occupancy model of T-cell activation is unable to explain MRL-related biological responses, and we propose here a hypothetical model of TCR-mediated signaling to accomodate the new data. In this model, TCR microaggregates whose organization has particular kinetic and spatial requirements represent the basic transduction unit of the signaling machinery. The quality and the quantity of the signals produced varies with the ligand, and the effects of the signals in the T cell are dependent on the interplay of both positive and negative downstream events whose effects relate to both the location and timing of their production. Further studies are clearly need to confirm or modify the details of this early model, but it seems clear that the discovery of

MRL has opened a new window into the molecular basis of T-cell activation events, provided new insight into the process of thymocyte development, and suggested new approaches to vaccine design and control of immune disorders.

References

1   Germain RN: The ins and outs of antigen processing and representation. Nature 1986;322: 687–689.
2   Yewdell JW, Bennink JR: The binary logic of antigen processing and presentation to T cells. Cell 1990;62:203–206.
3   Schwartz RH: Immune response (IR) genes of the major histocompatibility complex. Adv Immunol 1986;38:31–201.
4   Schwartz RH: Acquisition of immunologic self-tolerance. Cell 1989;57:1073–1081.
5   von Boehmer H: Thymic selection: A matter of life and death. Immunol Today 1992;13: 454–458.
6   Rammensee HG, Falk K, Rotzschke O: MHC molecules as peptide receptors. Curr Opin Immunol 1993;5:35–44.
7   Racioppi L, Ronchese F, Schwartz RH, Germain RN: The molecular basis of class II MHC allelic control of T cell responses. J Immunol 1991;147:3718–3727.
8   Hedrick SM, Matis LA, Hecht TT: The fine specificity of antigen and Ia determinant recognition by T cell hybridoma clones specific for PCC. Cell 1982;30:141–152.
9   Sette A, Buus S, Colon S: Structural characteristics of an antigen required for its interaction with Ia and recognition by T cells. Nature 1987;328:395–399.
10 Allen PM, Matsueda GR, Evans RJ: Identification of the T cell and Ia contact residues of a T cell antigenic epitope. Nature 1987;327:713–715.
11 Rothbard JB, Lechler RI, Howland K: Structural model of HLA-DR1 restricted T cell antigen recognition. Cell 1988;52:515–523.
12 Evavold BD, Williams SG, Hsu BL, Buus S, Allen PM: Complete dissection of the Hb (64–76) determinant using T-helper 1, T-helper 2 clones, and T-cell hybridomas. J Immunol 1992;148: 347–353.
13 Rothbard JB, and Gefter ML: Interactions between immunogenic peptides and MHC proteins. Annu Rev Immunol 1991;9:527–565.
14 Heldrick SM, Engfel I, McElliot DL, Finck PJ, Hsu ML, Hansburg D, Matis LA: Selection of amino acid sequences in the beta chain of the T cell antigen receptor. Science 1988;239: 1541–1544.
15 Acha-Orbea H, Mitchell DJ: Limited heterogeneity of T cell receptors from lymphocytes mediating autoimmune encephalomyelitis allows specific immune intervention. Cell 1988;54: 263–273.
16 Urban JL, Kumar V, Kono DH: Restricted use of T cell receptor V-genes in murine autoimmune encephalomyelitis raises possibilities for antibody therapy. Cell 1988;54:577–592.
17 Lai MZ, Jang YJ, Chen JK, and Gefter ML: Restricted V-(D)-J junctional regions in the T cell response to λ repressor. J Immunol 1990;144:4851–4856.
18 Sorger SB, Paterson Y, Fink PJ, Hedrick SM: T cell receptor junctional regions and the MHC molecule affect the recognition of the antigenic peptides by T cells clones. J Immunol 1990;144: 1127–1135.
19 Danska JS, Livingstone AM, Paragas V, Ishihara T, Fathman CG: The presumptive CDR3 regions of both T cell receptor α and β chains determine T cell specificity for myoglobin. J Exp Med 1990;172:27–33.
20 Jogensen JL, Fazekas de St. Groth B, Reay PA, Davis MM: Mapping T-cell receptor-peptide contacts by variant peptide immunization of single chain transgenics. Nature 1992;355:224–230.
21 Ehrich EW, Devaoux B, Rock EP, Jorgensen JL, Davis MM, Chien Y: T cell receptor interaction with peptide/MHC and superantigen/MHC ligands is dominated by antigen. JExp Med 1993;178:713–722.

22 Ashwell JD, Klausner RD: Genetic and mutational analysis of the T-cell antigen receptor. Annu Rev Immunol 1990;8:139–167.

23 Letourneur F, Klausner RD: T-cell and basophil activation through the cytoplasmic tail of T-cell-receptor zeta family proteins. Proc Natl Acad Sci USA 1991;88:8905–8909.

24 Wegener AM, Letourneur F, Hoeveler A, Brocker T, Luton F, Malissen B: The T cell receptor/CD3 complex is composed of at least two autonomous transduction modules. Cell 1992;68:83–95.

25 Samelson LE, Patel MD, Weissman AM, Harford JB, Klausner RD: Antigen activation of murine T cell induces tyrosine phosphorylation of a polypeptide associated with the T cell antigen receptor. Cell 1986;46:1083–1090.

26 Weiss A, Imboden J, Shoback D, Stobo J: Role of T3 surface molecules in human T-cell activation: T3-dependent activation results in an increase in cytoplasmic free calcium. Proc Natl Acad Sci USA 1984;81:4169–4173.

27 June CH, Fletcher MC, Ledbetter JA, Samelson LE: Increases in tyrosine phosphorylation are dectable before phospholipase C activation after T cell receptor stimulation. J Immunol 1990;144:1591–1599.

28 Kammer GM, Boehm CA, Rudolph SA, Schultz LA: Mobility of the human T lymphocyte surface molecules CD3, CD4, and CD8: Regulation by a cAMP dependent pathway. Proc Natl Acad Sci USA 1988;85:792–796.

29 Laxminarayana D, Berrada A, Kammer GM: Early events of human T lymphocyte activation are associated with type I protein kinase A activity. J Clin Invest 1993;92:2207–2214.

30 Weawer CT, Hawrylowicz CM, Unanue ER: T helper cell subsets require the expression of a distinct costimulatory signals by antigen-presenting cells. Proc Natl Acad Sci USA 1988;85:8181–8188.

31 Muller DL, Jenkins MK, Chiodetti L, Schwartz RH: An intracellular calcium increase and protein kinase C activation fail to initiate T cell proliferation in the absence of a costimulatory signal. J Immunol 1990;144:3701–3709.

32 Linsley PS, Brady W, Grosmaire L, Aruffo A, Damle NK, Ledbetter KJA: Binding of the B cell activation antigen B7 to CD28 costimulates T cell proliferation and interleukin 2 mRNA accumulation. J Exp Med 1991;173:721–730.

33 Koulova L, Clark EA, Shu G, Dupont B: The CD28 ligand B7/BB1 provides costimulatory signal for alloactivation of CD4+ T cells. J Exp Med 1991;173:759–762.

34 Vandenberghe P, Freeman GJ, Nadler LM, Fletcher MC, Kamoun M, Turka LA, Ledbetter JA, Thompson CB, June CH: Antibody and B7:BB1-mediated ligation of the CD28 receptor induces tyrosine phosphorylation in human T cells. J Exp Med 1992;175:951–960.

35 Jenkins MK, Pardoll DM, Mizuguchi J, Quill H, Schwartz RH: T-cell unresponsiveness in vivo and in vitro: Fine specificity of induction and molecular characterization of the unresponsive state. Immunol Rev 1987;95:113–182.

36 Bach FH, Grillot CC, Kupermann OJ, Sollinger HW, Hayes C, Sondel PM, Alter BJ, Bach ML: Antigenic requirements for triggering of cytotoxic T lymphocytes. Immunol Rev 1977;35:76–107.

37 Germain RN: Accessory cell stimulation of T cell proliferation requires active antigen processing, Ia-restricted antigen presentation, and a separate nonspecific second signal. J Immunol 1981;127:1964–1972.

38 Jenkins MK, Schwartz RH: Antigen presentation by chemically modified splenocytes induces antigen-specific T cell unresponsiveness in vitro and in vivo. J Exp Med 1987;165:302–319.

39 Mueller DL, Jenkins MK, Schwartz RH: An accessory cell-derived costimulatory signal acts independently of protein kinase C activation to allow T cell proliferation and prevent the induction of unresponsiveness. J Immunol 1989;142:2617–2628.

40 Azuma M, Ito D, Yagita KO, Phillips JH, Lanier LL, Somoza C: B70 antigen is a second ligand for CTLA-4 and CD28. Nature 1993;366:76–79.

41 Hathcock KS, Laszio G, Dikler HB, Bradshaw J, Linsley P, Hodes R: Identification of an alternative CTLA-4 ligand costimulatory for T cell activation. Science 1993;262:905–911.

42 Liu YB, Jones B, Aruffo A, Sullivan KM, Linsley PS, Janeway CA Jr: Heat-stable antigen is a costimulatory molecule for CD4 T cell growth. J Exp Med 1992;175:437–445.

43 Kay R, Takei F, Humphries RK: Expression cloning of a cDNA encoding M1/69J11d heat-stable antigens. J Immunol 1990;145:1952–1959.

44 Janeway CA, Bottomly K: Signals and signs for lymphocyte responses. Cell 1994;76:275–285.

45 Matis LA, Glimcher LH, Paul WE, Schwartz RH: Magnitude of response of histocompatibility-restricted T-cell clones is a function of the product of the concentrations of antigen and Ia molecules. Proc Natl Acad Sci USA 1983;80:6019–6028.

46 Ashwell JD, Fox BS, Schwartz RH: Functional analysis of the interaction of the antigen-specific T cell receptor with its ligands. J Immunol 1986;136:757–766.

47 Mannie MD: A unified model for T cell antigen recognition and thymic selection of T cell repertoire. J Theor Biol 1991;151:169–192.

48 Singer SJ: Intracellular communication and cell-cell adhesion. Science 1992;255:1671–1682.

49 Fiering S, Northrop JP, Nolan GP, Mattila PS, Crabtree GR, Herzenberg LA: Single cell assay of a transcription factor reveals a threshold in transcription activated by signals emanating from the T-cell antigen receptor. Genes Dev 1990;4:1823–1828.

50 Evavold BD, Allen P: Separation of IL-4 production from Th cell proliferation by an altered T cell receptor ligand. Science 1991;252:1308–1310.

51 Evavold BD, Sloan-Lancaster J, Hsu BL, Allen P: Separation of T helper 1 clone cytolysis from proliferation and lymphokine production using analog peptides. J Immunol 1993;150:3131–3140.

52 Sloan-Lancaster J, Evavold BD, Allen P: Induction of T cell anergy by altered T-cell-receptor ligand on live antigen-presenting cell. Nature 1993;363:156–159.

53 De Magistris MT, Alexander J, Coggeshall M, Altman A, Gaeta FCA, Grey HM, Sette A: antigen analog-major histocompatibility complexes acts as antagonists of the T-cell receptor. Cell 1992;68:625–634.

54 Alexander J, Snoke K, Ruppert J, Sydney J, Wall M, Southwood S, Oseroff C, Arrhenius T, Gaeta FCA, Colon SM, Grey HM, Sette A: Functional consequences of engagement of the T cell receptor by low affinity ligands. J Immunol 1993;150:1–7.

55 Racioppi L, Ronchese F, Matis LA, Germain RN: Peptide major histocompatibility complex class II complexes with mixed agonist-antagonist properties provide evidence for ligand differences in T cell receptor-dependent intracellular signaling. J Exp Med 1993;177:1047–1060.

56 Jamenson SC, Carbone FR, Bevan MJ: Clone-specific T cell receptor antagonists of major histocompatibility complex class I-restricted cytotoxic T cell. J Exp Med 1993;177:1541–1550.

57 Kronke M, Leonard WJ, Depper JM, Greene WC: Sequential expression of genes involved in human T lymphocyte growth and differentiation. J Exp Med 1985;161:1593–1607.

58 Weiss A, Littman DR: Signal transduction by lymphocyte antigen receptors. Cell 1994;76:263–274.

59 Skålhegg BS, Taskén K, Hansson V, Huitfeltd HS, Jahnsen T, Lea T: Location of cAMP-dependent protein kinase type I with the TCR-CD3 complex. Science 1993;263:84–87.

60 Ruppert J, Alexander J, Snoke K, Coggeshall M, Herbert E, Mc Kenzie D, Grey H, Sette A: Effect of T cell receptor antagonism on interaction between T-cells and antigen-presenting cells and on T-cell signaling events. Proc Natl Acad Sci USA 1993;90:2671–2675.

61 Brown TH, Jardetzky TS, Gorga JC, Stern LJ, Urban RG, Strominger JL, Wiley DC: Three-dimensional structure of the human class II histocompatibility antigen HLA-DR1. Nature 1993;364:33–39.

62 Dintzis HM, Dintzis RZ, Vogelstein B: Molecular determinants of immunogenicity: the immunon model of immune response. Proc Natl Acad Sci USA 1976;73:3671–3675.

63 Ashton-Rickardt PG, Van Kaer L, Schumacher TNM, Ploegh HL, Tonegawa S: Peptide contributes to the specificity of positive selection of CD8+ T cell in the thymus. Cell 1993;73:1041–1049.

64 Hogquist KA, Jamenson SC, Heath WR, Howard J, Bevan MJ, Carbone FR: T cell receptor antagonist peptides induce positive selection. Cell 1994;76:17–27.

65 Mossmann TR, Cherwinsky H, Bond MW, Giedlin MA, Coffmann RL: Two types of murine helper T cell clone. I. Definition according to profiles of lymphokines activities and secreted proteins. J Immunol 1986;136:2348–2357.

66   Swain SL, Bradley LM, Croft M, Tonkonogy S, Atkins G, Weinberg AD, Duncan DD, Hedrick SM, Dutton RW, Huston G: Helper T-cell subsets: Phenotype, function and role of lymphokines in regulating their development. Immunol Rev 1991;123:115–144.

67   Bloom R, Modlin RL, Salgame P: Stigma variations: Observation on suppressor T cell and leprosy. Annu Rev Immunol 1993;10:453–487.

68   Clerici M, Shearee GM: A $T_H1 \rightarrow T_H2$ switch is a critical step in the etiology of HIV infection. Immunol Today 1993;14:107–111.

69   Urban JJ, Katona IM, Paul WE, Finkelmann FD: Interleukin-4 is important in protective immunity to a gastrointestinal nematode infection in mice. Proc Natl Acad Sci USA 1991;88: 5513–5517.

70   Adorini L, Nagy ZA: Peptide competition for antigen presentation. Immunol Today 1990;11: 21–24.

71   Lanzavecchia A, Reid PA, Watts C: Irreversible association of peptides with class-II MHC molecules in living cells. Nature 1992;357:249–252.

Dr. Luigi Racioppi, Laboratorio di Immunologia, Dipartimento di Biologia e
Patologia Cellulare e Molecolare, Università di Napoli 'Federico II', Torre Biologica 14° pano,
Via Pansini 5, I–80131 Napoli (Italy)

Adorini L (ed): Selective Immunosuppression: Basic Concepts and Clinical Applications.
Chem Immunol. Basel, Karger, 1995, vol 60, pp 100–125

....

Idiotypic Regulation Directed at T-Cell Receptor Determinants[1]

Arthur A. Vandenbark [a–c], *George A. Hashim* [d], *Halina Offner* [a,c]

[a] Neuroimmunology Research 151D, V. A. Medical Center;
[b] Department of Molecular Microbiology and Immunology, Oregon Health Sciences University, and
[c] Department of Neurology, Oregon Health Sciences University, Portland, Oreg.;
[d] Council for Tobacco Research, New York, N. Y., USA

The major focus of our laboratories has been the characterization and regulation of the encephalitogenic process. To this end, we have employed both the rat and mouse models of experimental autoimmune encephalomyelitis (EAE), in which definitive studies of the phenotype, function, and migration patterns of encephalitogenic and regulatory T cells can be carried out. Simultaneously, we have carried out parallel studies in patients with multiple sclerosis (MS), in whom encephalitogenic T cells may contribute to the disease pathogenesis. Although it is not yet possible to definitively determine whether myelin autoreactive T cells cause MS, or even if they are encephalitogenic, it is informative to compare their properties with rodent T cells with demonstrated encephalitogenic activity. As will be presented below, there are remarkable similarities in the properties and frequencies of human and rodent myelin basic protein specific T cells that support the involvement of human T cells in MS. Similarly, we have extended what we have learned about immunoregulation directed at TCR peptides in rats and mice to the clinic, where a phase I treatment trial in MS patients has now been completed. Again, although the regulatory activity of TCR peptide specific T cells cannot be demonstrated directly in humans, there are extraordinary similarities of these T cells with rodent TCR peptide specific T cells that can transfer protection against EAE.

Rather than to simply recapitulate our prior published data, the primary goal of this review is to critically compare the known pathogenic and regula-

[1] This work was supported by the Department of Veterans Affairs, by NIH grants NS23221, NS23444, and by XOMA Corporation.

tory components in rodent EAE with their counterparts in MS. Although this sort of analysis cannot prove that MS is a form of EAE that can be treated with TCR peptides, it does serve to assess the degree of similarity between human and rodent immune responses. Thus, we have learned that there are a sufficient number of myelin basic protein specific T cells with properties similar to rodent encephalitogenic T cells to account for the immunopathology observed in MS, at least in some patients. This conclusion forms the rationale to regulate BP-reactive T cells to determine their contribution to MS. Towards this goal, we have established that immunity directed at TCR peptides exists in humans, and can be boosted to a level sufficient for regulation. Thus, although many additional problems must be overcome, idiotypic regulation directed at TCR determinants remains feasible as an approach for selective intervention in human autoimmune disease.

The Pathogenic Processes in EAE and MS

Clinical and Histopathological Features

EAE is a T-cell-mediated disease directed at central nervous system (CNS) myelin proteins, including basic protein (BP) [1], proteolipid protein (PLP) [2–4], and possibly other myelin components. Both BP and PLP are widely encephalitogenic among mammalian species and strains when injected with adjuvants, inducing clinical and histological sequelae ranging from self-limiting paralysis without obvious demyelination (e.g. Lewis and LOU M rats) to relapsing disease with extensive demyelination similar to MS (e.g. PL/J and SJL/J mice, strain 13 guinea pigs, monkeys). Lesions are perivascular and contain mononuclear cells, and usually predominate in the lower spinal cord, thus accounting for ascending paralysis characteristic of EAE. Some rodent strains are relatively resistant to EAE (e.g. BALB/c mice, Brown Norway rat), requiring additional immune potentiation (pertussogen) to induce clinical signs in only a fraction of the injected animals. Thus, the genetic background, including MHC class II alleles, strongly influences the ease of induction and the clinical course of EAE, as well as the degree of demyelination.

Two studies from our laboratory illustrate this point. The first involved transferring BP- or PLP-reactive T cells from Lewis rats into severe combined immunodeficient (SCID) mice (H-2^b) that had been reconstituted with LEW hemopoietic cells to provide histocompatible APC [5]. LEW T cells specific for a PLP peptide (residues 139–151 known to be highly encephalitogenic in SJL/J mice [2]) were unable to induce clinical or histological disease in LEW rats. However, when transferred into the SCID/LEW mice, these T cells induced paralytic EAE with demyelination. Moreover, LEW BP-specific T cells that caused EAE lasting 7 days with no demyelination in LEW rats induced EAE lasting 21 days with extensive demyelination in SCID/LEW mice. The more extensive clinical course may have resulted from regulatory elements present in LEW rats but absent in SCID/LEW mice that normally prevent or limit the activity of the LEW T cells, or alternatively, the SCID mouse may possess crucial differences that enhance EAE expression. Moreover, it was apparent that the SCID mice, rather than either LEW rat APC or BP-specific T cells, contributed crucial determinants permissive for the expression of cell-mediated demyelination. In the second example, we found that BUF rats with EAE developed extensive demyelination, whereas LEW rats did not, even though the severity of EAE and the number of CNS infiltrating cells were the same [6]. Moreover, (LEW × BUF)F1 rats did not develop demyelination, even when EAE was induced by transferred BP-specific BUF T cells [Jones, personal commun.]. F1 BP-specific T cells transferred EAE to sublethally irradiated BUF parents, but did not cause demyelination. The only combination that produced EAE with demyelination was BUF T cells to BUF recipients. Thus, in this instance, demyelination was determined by both the source of encephalitogenic T cells and the target CNS tissue.

A similar type of T-cell-mediated inflammation directed at CNS components would likely produce a spectrum of clinical signs and histological lesions in the outbred human population. Thus, acute disseminated encephalomyelitis (ADE) might represent the human form of monophasic EAE, whereas MS, which is marked by progressive CNS damage, perivascular lesions, and demyelination, would be more similar to chronic relapsing EAE. One prominent feature in MS that has not been observed in CR/EAE is the formation of plaques, which are fibrin deposits resulting from active gliosis in areas of extensive demyelination [7]. Whether plaque formation is a unique function of human glial cells or results from long-term chronic CNS inflammation remains to be determined. It is conceivable that demyelination is a prerequisite for plaque formation, and it may be that the MS disease process leading to plaque formation can only occur in individuals permissive for demyelination. It should be noted that relapsing neurological disease with demyelination does not require myelin-specific T cells. In mice infected with Theiler's murine encephalitis virus (TMEV), relapses with 'bystander' demyelination are initiated by T cells specific for viral antigens that appear to be distinct from any of the known myelin encephalitogens [8, 9]. Later episodes, however, might result from the subsequent induction of myelin antigen-specific T cells in the damaged tissue [Miller, personal commun.] by a process termed 'determinant spreading' [10]. Similarly, MS could be initiated by inflammation directed at an infectious agent in the CNS, resulting in the induction of myelin antigen-specific T cells that drive progressive disease.

Table 1. Estimated frequencies (cells/million) of BP-specific T cells in EAE and MS

	EAE		MS	
	baseline	disease onset	baseline	maximum
Blood	≤1	11	1	7
CNS/CSF	<10	500	10	220

Characteristics of Encephalitogenic T Cells

Frequency

In our studies in rats, we found that upon immunization, the relative frequency (measured by short-term proliferation) of BP-specific T cells increased from a baseline of ≤1 cell/million blood or lymph node cells to >11 cells/million at onset of EAE [11] (table 1). In the CNS tissue, the relative frequency was much higher, reaching >500 cells/million just prior to onset of clinical signs (table 1). Others, using more sensitive proliferation-based assays, found even higher relative frequencies than we reported [12], although the pattern of increased frequency in CNS versus the periphery has been a consistent finding.

It is now clear that BP-reactive T cells can be found in the circulation of all humans, including those with no evidence of neurological disease. Using the same proliferation-based assay as in animals, we established the estimated frequency of BP-reactive T cells in the blood and CSF of MS and control patients [13]. In our initial study, we found that the frequency in MS patients was, on average, about 6 cells/million (range 2–14 cells/million), approximately half of that observed in rats with EAE. Patients with other neurological diseases (OND), rheumatoid arthritis, or with no disease were significantly lower, ranging from 1 to 2 cells/million. In a follow-up longitudinal study [14], we found that the estimated frequency of BP-specific T cells in MS patients was episodic, increasing from a baseline value of 1–2 cells/million to a peak value of about 7 cells/million for a period of several months, and then declining back to baseline (fig. 1). During this period of increased BP frequency, the MS patients suffered more relapses than during periods of baseline responses, although significant clinical changes were not observed in several of the patients in whom the BP-specific T-cell frequency increased. These data suggested that some but not all of the surges in BP-specific T cells may have measurable clinical consequences, consistent with MRI data [15].

In the CSF, we estimated the frequency of activated (and perhaps more disease-relevant) BP-specific T cells after expansion in IL-2/IL-4 [13]. In MS

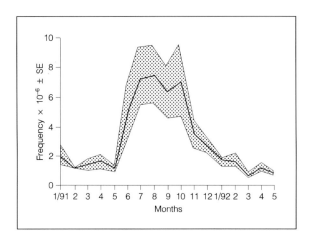

Fig. 1. Composite longitudinal BP-specific T-cell frequency of 12 MS patients. For each determination, blood mononuclear cells were separated on a ficoll density gradient and 10–24 replicate wells containing 4 twofold dilutions of 500,000 cells/well were cultured with APC in the presence or absence of human BP for 5 days prior to harvesting and evaluation of proliferation (^{3}H-Tdy uptake). In BP-reactive wells, the cpm exceeded 2 SD of the mean cpm of wells without BP. For each patient/month, the number of negative wells at each cell dilution was used to estimate the frequency ± 95% confidence interval as described [14]. The same BP preparation was used throughout the study. From Vandenbark et al. [14].

patients, the estimated frequency of BP-reactive T cells was 220 cells/million (range 56–400 cells/million), again approximately half of that found in the CNS of rats developing EAE and significantly higher than in OND patients, who had 12 cells/million (range 1–40 cells/million). Although the frequencies of BP-reactive T cells are similar in the CNS of rats with EAE and in the CSF of patients with MS, there is an important difference in cell concentrations. Thus, rat CSF, which is largely reflective of CNS T-cell specificities [16], contains a much higher concentration of cells (about 800 cells/μl) [17] than the CSF of MS patients (0.5–15 cells/μl) [13]. Because the concentration difference is partially offset by total CSF volume (≤1 ml in the rat versus >100 ml in humans), the total number of cells in CSF would appear to be similar in rats and humans (roughly 1/2 million total cells each). The question remains as to whether or not this total cell load of BP-specific T cells is sufficient to account for clinical signs in MS.

Of additional interest, MS and OND patients had similar estimated frequencies (13 and 16 cells/million, respectively) of T cells specific for 'cryptic' epitopes of BP (e.g. response to BP fragments but not to intact BP) [18]. In

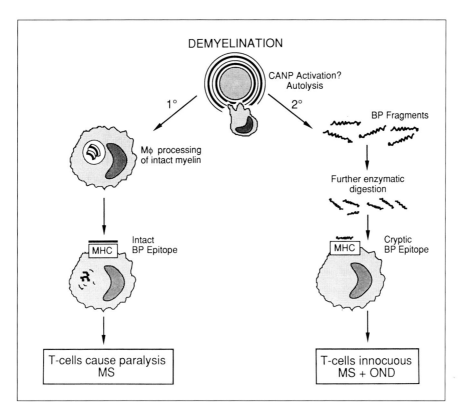

Fig. 2. Proposed antigen-processing pathways that distinguish induction of whole BP-reactive T cells characteristic of MS patients from induction of BP fragment-reactive T cells characteristic of both MS and OND patients. The key difference in the two pathways is that primary demyelination involves direct stripping and processing of BP by macrophages or microglial cells leading to the preservation of intact BP epitopes, whereas BP released during secondary demyelination will be further degraded by autocatalytic enzymes into 'cryptic' epitopes that differ antigenically from the intact BP epitopes.

that responses to 'cryptic' epitopes were not encephalitogenic in rats [19], these data suggest that there may be two distinct pathways for processing BP in CNS (fig. 2). One pathway would involve intact BP as might occur when myelin is stripped and engulfed by microglial cells or macrophages during primary demyelination in MS plaques, and would give rise to the majority of BP-reactive clones in MS CSF. The second pathway would involve unusual BP fragments not typical to processed BP, such as those resulting from autocatalytic or extracellular degradation of soluble BP [20], and would give rise mostly to peptide-specific, variant T cells clones. Our results suggest that the first pathway may be related to the MS disease process, whereas the second, present in both MS and OND patients, may result from other types of CNS damage. This conclusion will obviously remain speculative until the encephalitogenic activity of human T cell clones can be established (see below).

Phenotype and Specificity

T cells that can transfer EAE are uniformly MHC class II restricted, CD4+ Th1-like memory cells (CD45RO) that produce inflammatory lymphokines including IFNγ, IL-2, and TNF upon activation with their specific CNS peptide ligand [21]. To date, BP and PLP are the only known CNS encephalitogens, although it seems likely that other CNS proteins could also induce encephalitogenic T cells. The specificity of encephalitogenic BP- and to a lesser degree PLP-specific T cells has been established in many species and strains [1], and it is clear that the encephalitogenic epitopes vary according to the MHC class II restricting element. There appear to be a limited number (10–12) of discrete encephalitogenic regions within BP that are used repeatedly on different genetic backgrounds [22]. In a given strain, only one or two epitopes are encephalitogenic. The focused T-cell response against encephalitogenic epitopes tends to be oligoclonal, resulting in T cells expressing the same Vα and Vβ genes, often with characteristic CDR3 motifs that have been postulated to interact directly with the encephalitogenic ligand.

We have characterized extensively the oligoclonal T-cell responses to two encephalitogenic epitopes in the LEW rat (table 2). When immunized with whole guinea pig BP (Gp-BP), the immunodominant response is directed at the 72–89 sequence restricted by I-A [23, 24]. The 72–89 specific T cells are strongly biased in their expression of Vα2 and Vβ8.2 [25], which in a majority of clones contains a characteristic ASP-SER CDR3 motif [26, 27]. However, upon prolonged immunization with Gp-BP [28], after immunization with bovine or human BP in which the 72–89 sequence is substantially altered [29, 30], after rendering the rats unresponsive to the 72–89 peptide by neonatal tolerization [31], or by direct immunization with the peptide [32], the response is enhanced or redirected to a secondary I-E restricted epitope, residues 87–99. T cells specific for 87–99 tend to utilize Vβ6 that contains a characteristic ARG-GLY CDR3 motif [33]. The occurrence of encephalitogenic specificities is greatly enhanced within the CNS tissue [34], but recovery of these cells is impaired by the low cellularity after resolution of clinical disease [17]. Neonatal tolerization with both the 72–89 and 87–99 epitopes prevents the subsequent induction of EAE with Gp-BP, although other nonencephalitogenic determinants are recognized [31]. This situation is similar to that in rats recovered from EAE, in which prolonged immunization to whole BP continues after responses to the encephalitogenic epitopes have diminished, resulting in a sub-

Table 2. Summary of TCR peptides tested in EAE and MS

Strain/species	BP or PLP-specific V gene bias	TCR peptide injected	
		effective	not effective
LEW rat	Vβ8.2 (BP 72–89)	Vβ8.2 9-59	Vβ8.2 25-41
		Vβ8.6 39-59	Vβ8.2 93-101
		Vβ8 44-54[1]	Vβ13 39-59
		Vβ8 44-52[2]	Vβ14 25-41
		Vβ6 36-58[2]	Vβ14 39-59
	Vβ6 (BP 87–99)	Vβ6 36-58	Vβ14 39-59
		Vβ8.2 39-59	
		Vβ8 44-54	
(LEW × BUF)F1 rat	Vβ8.2 (BP 72–89)	Vβ8.2 39-59	Vβ14 39-59
SJL/J mouse	Vβ2, 4,10, 16, 17a (PLP 139–151)	Vβ4 42-63 & Vβ17a 1-17	Vβ17a 47-66 Vβ17a 51-66
(PL × SJL)F1 mouse	Vβ4, 8.2 (PLP 43–64)	Vβ4 42-63 & Vβ8.2 39-62	
Human	Vβ5.2, 6.1 (BP)	Vβ5.2 39-59 Vβ6.1 39-59	

[1] Consensus peptide of Vβ8.2 39-59 and Vβ8.6 39-59.
[2] Cross reactive with Vβ8.2 39-59.

stantial increase of nonencephalitogenic T-cell specificities, including those recognizing an EAE-protective epitope, residues 55–67 [35, 36]. These findings demonstrate that the encephalitogenic activity of BP in LEW rats is limited to only two epitopes. However, as a prelude to human studies, it is important to note that because the response to encephalitogenic epitopes is obfuscated during recovery, it is virtually impossible to identify the pathogenic T cell specificities without transferring passive EAE. A further consideration is that encephalitogenic T cells must recognize epitopes that are retained after processing of the intact BP present in the CNS of recipient animals. 'Cryptic' epitopes that do not represent naturally processed regions of BP are not encephalitogenic [19], supporting the notion that the pathogenic T cells must be restimulated within the CNS for EAE to occur.

Like encephalitogenic T cells in rodents, BP-responsive T cells from humans are predominantly MHC class II restricted CD4+ Th1-like memory cells [37–39] that produce inflammatory lymphokines, although in rare instances, class I-restricted CD8+ T cells can be isolated that have regulatory functions [40]. As would be expected in an outbred population, the BP epitope specificity of human T cells is relatively heterogeneous. As a group, humans recognize essentially the same set of 10–12 epitopes on BP as the collection of inbred animal T cells [38, 39]. Individually, however, T cell responses to BP are more limited, usually to one or two dominant epitopes in normal donors. In MS patients, the response to BP often involves additional specificities, suggesting sensitization to minor epitopes during the disease process [38] reminiscent of rats recovered from EAE. As is discussed in another review article [41], we found roughly 1/3 of blood-derived and 2/3 of CSF-derived BP-specific clones responded to epitopes in the N-terminal half of BP, with less pronounced responses to the 84–102 and 139–151 epitopes than observed by others. Responses to residues 45–89, 84–102, and 139–149 were observed more frequently in MS patients than controls, but it remains to be seen if T cells responsive to these epitopes are encephalitogenic. Nearly all (>90%) of the responses to BP were restricted by HLA-DR alleles [38]. HLA-DR, which is a risk factor in MS, could act as a restriction molecule for virtually the entire spectrum of BP epitopes, suggesting that this allele may be permissive for many encephalitogenic determinants.

V gene use. It is now well established that encephalitogenic T cells from a number of rodent strains tend to utilize a limited number of TCR Vα and Vβ genes in response to BP, notably Vα2 and Vβ8.2 [42]. In the LEW rat, we found that the most pronounced expression of Vβ8.2 occurred in the CNS just prior to disease onset, and that expression of Vβ8.2 could be enhanced by culturing CNS or CSF T cells in IL-2, or by restimulating the extracted T-cell populations with BP [17]. The importance of these particular V genes has been demonstrated unequivocally in that double transgenic mice which overexpress encephalitogenic Vα2 and Vβ8.2 sequences in their TCR develop spontaneous EAE [43]. In other strains such as the SJL/J mouse, in which nearly half of the Vβ genes including Vβ8.2 have been deleted, the response to BP or PLP peptide is less biased, involving predominantly Vβ2, Vβ4, Vβ10, Vβ16 and Vβ17a [44]. The underlying cause of V gene bias in response to BP or other antigens remains unknown, but could involve preferential interactions of CDR1 and CDR2 with MHC molecules [45–47], as well as superantigen driven expansion of selected Vβ genes [48].

In humans, there are many examples of V gene bias in autoimmune diseases, cancer, and transplantation. However, there are many inconsistencies in data gathered by different investigative teams on patients with the same disease but from different areas or sampled at different times. In MS, there is substantial disagreement among at least four studies that have assessed V gene bias in BP-reactive T cells [49–52]. These discrepancies may be accounted for in part by differences in T-cell isolation techniques, periods of elevated BP-specific T-cell frequencies, MHC restriction, variation in clinical status of the patients, and physical location and thus exposure to superantigens. We have carried out three separate and overlapping studies to identify biased V gene expression in MS patients and controls. In our initial study, we found a substantial bias of Vβ5.2 and Vβ6.1 among BP-specific T cells from the blood of MS patients [51]. This finding was strengthened by the concurrent demonstration that T cells within demyelinating areas of MS brain plaques from DR2/Dw2 cadavers preferentially expressed the same two Vβ genes and had CDR3 motifs characteristic of BP-specific T cells [53]. In a second study, we found biased expression of Vα2, Vβ7, and Vβ18 among IL-2/IL-4 expanded BP-reactive T-cell clones isolated from the CSF of MS patients [18]. In the third study, we evaluated Vβ gene expression in bulk-cultured CSF T cells (not clones) that were first expanded with IL-2/IL-4, PHA, or anti-CD3, and then screened for BP reactivity [54]. This study demonstrated substantial differences in the nature and V gene use in the respective expanded T-cell populations. Of importance, BP reactivity was detected most often in the IL-2/IL-4 expanded population, and involved expression of Vβ1, Vβ2, Vβ5, and Vβ18. However, the IL-2/IL-4 expanded populations did not always contain the same subpopulations as BP-selected T cells. Moreover, simultaneous comparisons of blood and CSF-derived T-cell populations suggested that the reactivity to BP in CSF provided the best reflection of clinical activity, although in at least one instance, the predominant Vβ gene associated with BP reactivity changed over time. In comparison, blood T cells provided a partial but longer-lasting reflection of the CSF BP reactivity and Vβ gene bias.

In conclusion, it would appear from the above discussion that with the possible exception of Vβ5.2 in DR2/Dw2 patients, there is no single Vβ gene that is used consistently by BP-reactive T cells in all MS patients. Collectively, the studies on V gene expression indicate that there are substantial biases characteristic for each patient and to a limited degree for the larger MS population within a given area and time. Optimally, screening for BP response is most reliable in the CSF and blood during periods of clinical activity, although analysis of blood responses in early remission may still provide some useful information.

Inflammatory Lymphokines. One approach for defining important features of encephalitogenic T cells is to identify lymphokines that are selectively released within the CNS. Thus, we developed a passive transfer model in which encephalitogenic LEW T cells were injected into (LEW × BUF)F1 rats, or vice versa, and their fate and function followed using the allelic RT7.1 (LEW) and RT7.2 (BUF) markers [21]. We found that approximately 40% of the T cells infiltrating the CNS at onset of EAE were of donor origin, with a subsequent

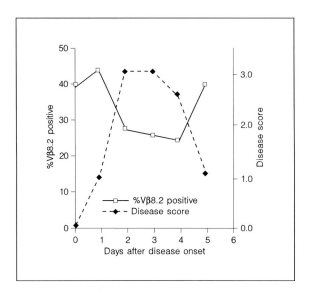

Fig. 3. Encephalitogenic activity and infiltration into spinal cord (SC) of LEW (RT7.2⁻) T cells transferred into (LEW × BUF)F_1 rats. SC cells were collected during the course of EAE in F_1 rats and double stained with RT7.2 and Vβ8.2 antibodies. From Weinberg et al. [21].

decline as host-recruited T cells entered the CNS during recovery (fig. 3). This experiment demonstrated unequivocally that encephalitogenic specificities provide the initial inflammatory phalanx. We further evaluated the lymphokine profiles of injected encephalitogenic T cells recovered from the CNS versus spleen, and found the selective upregulation of mRNA for IFNγ, IL-2, and IL-3 in CNS (fig. 4). Moreover, within the CNS, mRNAs for the above inflammatory lymphokines and for TNF were produced predominantly by the donor encephalitogenic T cells rather than by host-recruited T cells.

Fig. 4. Lymphokine mRNA production by transferred encephalitogenic T cells differs in the spinal cord (SC) versus the spleen (Sp) of Lewis rats with EAE. SC and Sp populations were stained with RT7.2-FITC and sorted after transfer of a BP-specific T-cell line from (LEW × BUF)F_1 (RT7.1⁺/RT7.2⁺) donors into irradiated LEW (RT7.2⁺/RT7.2⁻) rats. RNA was isolated from the two RT7.2⁺ post-sort populations which were 98 and 99% pure for Sp and SC cells, respectively. The samples were normalized with the GAPDH primers and the same amount of cDNA was added for the lymphokine primers. The DNA was visualized by EtBr staining of agarose gels. Note the increased expression in SC of IFNγ, IL-2, and IL-3 by the transferred encephalitogenic F_1 T cells. From Weinberg et al. [21].

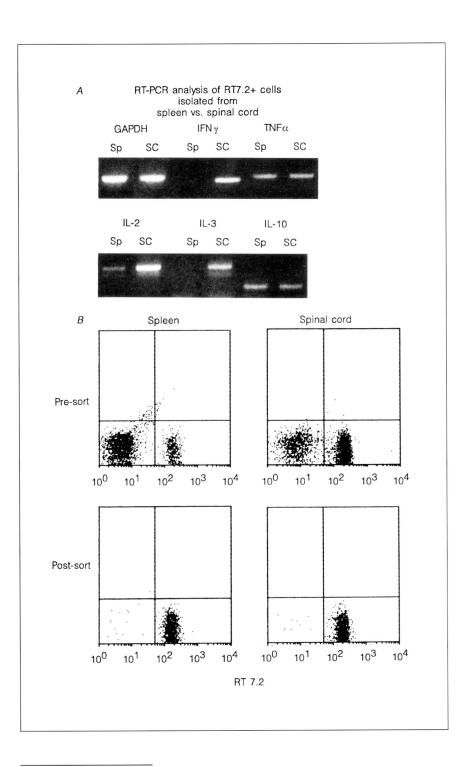

Similarly, human BP-reactive T-cell lines, upon activation with BP or mitogens, produced message for inflammatory lymphokines, including IFNγ and GM-CSF, with virtually no production of IL-4 and IL-5 [55], suggesting a Th1 subtype.

Encephalitogenic Activity. Identifying pathogenic cells in EAE is relatively straightforward through passive transfer experiments. A key issue is to identify which, if any, T cells actually contribute to the pathogenesis of human diseases such as MS. Towards this goal, we have begun to develop a chimeric SCID mouse model (C.B-17-*scid/scid*, H-2^d) in which the mice are reconstituted with bone marrow cells from allogeneic or xenogeneic animals to provide APC function that is histocompatible for transferred encephalitogenic T cells [5]. Thus far, we have successfully transferred clinical EAE with demyelination from SJL/J (H-2^s) mice, LEW rats, and BUF rats into reconstituted SCIDs. Currently, experiments are in progress to reconstitute the SCID mice with human bone marrow and then to transfer human T cells reactive to mouse BP [56] to assess encephalitogenic activity. A different approach, published by others, has demonstrated that CSF cells from MS patients with early active disease and elevated CSF cellularity could induce paralysis in SCID mice after intrathecal injection [57]. Using the identical protocol, we have been unable to induce paralysis or CNS lesions in SCID mice using MS CSF cells, human BP-specific T cells, or rat encephalitogenic BP-specific T cells injected intrathecally with or without additional APC [unpubl.]. The only source of cells that induced neurological disease after intrathecal injection was with unfractionated splenocytes, implicating a graft-vs.-host reaction. From these data, we conclude that the encephalitogenic process requires passage of suitably activated T cells across the blood-brain barrier into the CNS. Further development of the xenogeneic transfer system will be invaluable in distinguishing potentially pathogenic specificities and their characteristics, including V gene bias, from nonencephalitogenic specificities similar to those observed in recovered rats.

Immunoregulation Using TCR Peptides

The biased expression of common germline V gene sequences by autoreactive T cells could provide a potential target for regulatory antibodies and T cells. Anti-idiotypic antibodies have been described to conformational determinants on the TCR. However, TCR specific T cells would require an MHC-restricted linear peptide determinant for recognition. If a portion of the Vα or Vβ chain polypeptides were internally processed in a manner similar to other antigenic proteins, then some component peptides could be bound and expressed on the T-cell surface in association with MHC I or II molecules. An MHC-associated TCR peptide would thus constitute a legitimate target structure for interaction with potentially regulatory T cells that may have escaped negative selection in the thymus. As will be discussed below, TCR peptide specific T cells and antibodies do indeed exist in both animals and humans, possessing potent regulatory effects on pernicious autoreactive T cells.

Choice and Administration of TCR Peptides

The almost exclusive use of Vβ8.2 by encephalitogenic BP-reative T cells in LEW rats provided an optimal situation for testing the potential regulatory effects of TCR peptides. Based on its predicted antigenic characteristics [58, 59] and inclusion of the exposed CDR2 loop, we chose the rat Vβ8.2 39-59 sequence (DMGHGLRLIHYSYDVNSTEKG) for our initial studies. This peptide was highly antigenic in rats, inducing DTH reactivity, as well as specific T cells and antibodies [60, 61]. Moreover, when injected subcutaneously at doses of 100–400 μg in CFA 30 days prior to a subsequent challenge with BP, the Vβ8.2 39-59 peptide induced virtually complete protection against EAE [60], a finding confirmed by Stevens et al. [62], and corroborated independently by Howell et al. [63] using a different peptide, but not by others [64–67]. When injected at the same time or 7 or 11 days after challenge with BP (prior to onset of clinical signs), the Vβ8.2 39-59 peptide reduced the severity and duration of EAE [41]. Surprisingly, we noted little difference in the EAE-suppressive effects when injecting the peptide subcutaneously in CFA or intradermally in saline. Moreover, we found that intradermal injection of the peptide on the first day of clinical signs of EAE could significantly reduce the maximum severity and shorten the duration of disease [68].

The optimal treatment effect required 100–200 μg TCR peptide, given subcutaneously, intradermally, or intraperitoneally, but not intravenously or orally, with a lesser effect achieved with doses as low as 10 μg peptide or as high a 1,000 μg peptide [69]. In most cases, only one injection of Vβ8.2 39-59 was sufficient to alter the course of EAE. However, in (PL×SJL)F1 mice, we found that injection of 100 μg of two relevant TCR CDR2 peptides every 4 days was more effective than a single dose of the peptides [44]. The enhanced treatment effect of periodic administration of low doses of peptide was a key factor in choosing a treatment regime for MS patients, which now involves 4 weekly intradermal injections of 100 μg of the TCR peptide in buffer diluent, followed by monthly injections.

To date, we have tested a more than 15 different TCR peptides in four strains of rats and mice, and in humans (table 2). Generally, antigenic CDR2 peptides have been the most successful at inhibiting EAE, the only exception being Vβ17a 1-17. The Vβ8.2 CDR1 peptide was not antigenic in LEW rats, and the CDR3 peptide was relatively inactive, in contrast to another report [63]. Both human Vβ5.2 39-59 and Vβ6.1 39-59 CDR2 peptides were strongly antigenic in most patients tested [70]. Currently, we are in the process of testing overlapping peptides for the complete rat Vβ8.2 and human Vβ5.2 sequences to determine which other regions are antigenic or EAE protective (in the case of Vβ8.2).

Biological Effects in vivo

The induction of specific DTH responses to TCR peptides suggested that the regulation of EAE was immunologically mediated [60]. We thus sought to determine what were the effects of peptide administration on the induction of TCR-specific T cells and antibodies, and on the frequency and function of BP-reactive T cells.

Vaccination. Rats pretreated with Vβ8.2 39-59 or Vβ8 44-54 peptides were virtually completely protected against clinical EAE [60, 71], but histological lesions eventually developed [41]. Assessment of infiltrating T cells revealed a drastic decrease in total and Vβ8.2+ T cells in the CNS of protected rats versus controls at the time of disease onset (fig. 5) [72]. Moreover, there was a significantly reduced number of infiltrating CD45RC^{hi} T cells that are associated with recovery of diseased rats [73]. Assumably, these cells were not recruited or needed in the recovery process, since TCR peptide immunization prevented disease onset. Interestingly, late in disease in control rats, there was a reduced but similar number of total and Vβ8.2+ T cells in the CNS of both recovering and protected rats (fig. 5). Surprisingly, Vβ8.2+ BP-reactive T cells that transferred EAE could still be isolated from the LN of protected rats, suggesting that vaccination with Vβ8.2 39-59 did not result in the deletion of encephalitogenic T cells.

Therapy. Treatment of rats on the first day of clinical signs of EAE with Vβ8.2 39-59 or Vβ8 44-54 peptides provides a unique situation for testing the regulatory effects of anti-TCR peptide immunity on encephalitogenic T cells that have already entered the CNS tissue. Clinically, the TCR peptide-treated rats develop less-severe disease with a shorter duration [68]. The clinical effect is quite rapid, being distinguishable within 24–48 h. We postulated that the rapid response was due to a boosting effect of an underlying immunity to the TCR peptide that was induced naturally as a consequence of the focused expansion of encephalitogenic Vβ8.2+ T cells during EAE induction.

Indeed, we detected significant changes induced by TCR peptide therapy in DTH reactivity and in the frequencies of both TCR peptide and BP-specific T cells. Rats that had no prior exposure to synthetic TCR peptides had significantly increased peptide-specific DTH and proliferation activity when treated with TCR peptide injected intradermally [41, 68]. This response was reflected

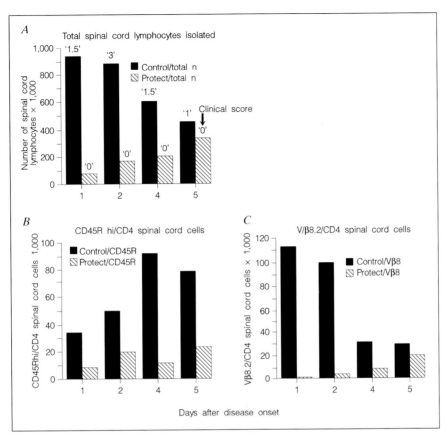

Fig. 5. Phenotype evaluation of spinal cord lymphocytes isolated from rats with EAE (control) vs. Vβ8.2 peptide-protected rats. The total number of SC lymphocytes isolated from control animals with EAE was compared to Vβ8.2-protected rats on days 1, 2, 4, and 5 after disease onset *(A)*. The control animals had clinical disease scores of 1.5, 3.0, 1.5 and 1.0, respectively, while protected animals showed no signs of disease. *B* The total number of CD45R^hi/CD4+ T cells isolated from spinal cords was compared between control and protected rats on days 1, 2, 4 and 5 after disease onset. *C* Comparison of the number of V 8.2+/CD4+ T cells isolated between control and protected animals. From Weinberg et al. [72].

by an increase in the frequency of TCR peptide-specific T cells in the blood and especially in the CNS where regulation presumably occurs [11]. Concomitantly, the frequency of BP-reactive T cells decreased significantly in both the periphery and the CNS. The decrease in BP reactivity was highly exaggerated within T cells specific for the major encephalitogenic epitope, residues 72–89 of BP, but this change was partially offset by an increased frequency of T cells specific for the protective 55–69 epitope of BP, especially in the draining lymph nodes [11]. These results indicated that regulation induced by TCR peptides altered the immunization process to BP, causing 'epitope switching'. Similar results were obtained in a prior study in which BP-specific T cell lines selected from animals protected by passive transfer of TCR-specific T cells had reduced responses to BP 72–89, but increased recognition of the 43–68 and other nonencephalitogenic regions of BP [60]. Although the response to BP was diminished in the CNS, it was apparent that there was no deletion of Vβ8.2+ T cells in TCR-peptide-treated rats, since Vβ8.2+ T cells were clearly detected in the CNS by fluorescent staining and by expression Vβ8.2 mRNA [72]. Moreover, there was no evidence of apoptosis in the CNS-derived T-cell population [unpubl. data].

Therapeutic injection of Vβ8.2 39-59 but not Vβ8 44-54 peptide also boosted production of peptide specific antibodies [74] consisting mainly of IgG and IgM [61]. Such anti-TCR peptide antibodies weakly stained Vβ8.2+ T cells, and could drastically reduce the severity of actively induced EAE. The ability of the Vβ8.2 39-59 peptide to boost antibody responses in animals not previously exposed to the synthetic peptide in vivo provides additional evidence that this sequence represents a naturally recognized idiotope.

The response pattern to TCR peptides was remarkably similar in 7 of 11 MS patients treated with Vβ5.2 39-59 and Vβ6.1 39-59 [70], 3 of whom were known to overexpress Vβ5.2 or Vβ6.1 in their BP-specific T-cell response [51]. Weekly injection of 100 μg TCR peptide induced a significant increase in the frequency of TCR peptide-specific T cells, usually within 1–3 weeks (fig. 6), and in the most vigorous responders, DTH responses were easily detected [70]. Nonresponders had no changes in T-cell frequency, even when doses as high as 3 mg were injected. Some patients who responded to lower doses of peptide exhibited decreased frequencies at doses ≥ 600 μg, consistent with the dose-response pattern seen in LEW rats. Anti-TCR peptide antibodies, including primarily IgG and IgM, were detected in two patients that also had increased T-cell frequencies after boosting with the TCR peptides.

The effect of TCR peptide therapy on BP response was complicated by the low frequency of BP-reactive T cells at the initiation of therapy in most patients, and by the episodic nature of the BP response itself (fig. 1). However, in a few cases, as illustrated in figure 7, BP responses clearly decreased as TCR peptide responses increased. In patient N. L., who had a strongly biased Vβ5.2 response to BP (13/14 clones), treatment with the illogical Vβ6.1 39-59 peptide had no effect on BP frequency, whereas treatment with the Vβ5.2 39-59 peptide decreased the BP frequency to background. Thus, although the changes were not as rapid as was observed in rats with acute EAE,

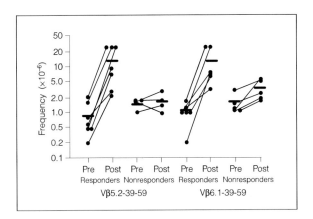

Fig. 6. Preimmunization and maximum postimmunization T-cell frequencies for Vβ5.2 39-59 and Vβ6 1 39-59 peptides are given for each patient. Patients were classified as responders or nonresponders based on a significant ($p \leq 0.05$) change in frequency above baseline value. The line indicates the arithmetic means of each group. From Bourdette et al. [70].

the injection of TCR peptides appeared to have a similar regulatory effect of reducing the frequency of BP-reactive T cells. Although the number of patients studied was too small to draw conclusions about the clinical effects of TCR peptide therapy in MS, there was a hint of efficacy in that 5 of 7 responders were clinically stable or better after 1 year, whereas only 1 of 4 nonresponders was stable.

Characterization of TCR Peptide-Reactive T Cells. Upon immunization with the Vβ8.2 39-59 or Vβ8 44-54 peptides in adjuvant, it was possible to select peptide-specific T-cell lines and clones from LEW rats that could efficiently transfer protection against EAE [60, 71]. The protective T cells were largely CD4+CD8^{dim}, and their responses to peptide presented by APC were inhibited predominantly by antibodies to MHC I and CD8 molecules, but not by antibodies to MHC II or CD4, suggesting MHC I restriction and involvement of the weakly co-expressed CD8 molecule. Additionally, the TCR peptide-reactive T cells could be stimulated by attenuated Vβ8.2+ T cells in the absence of APC, indicating the likelihood of an immunogenic MHC/TCR antigen complex on the stimulator T-cell surface. The regulatory T cells expressed heterogeneous TCR Vβ genes, including Vβ4, 8, 10, 12, 15, 17, 19, and 20, unlike the focused Vβ8.2 gene expression observed in response to BP epitopes described above.

TCR peptide-specific T cells could be obtained with relative ease prior to peptide injection from patient N. L. (with biased Vβ5.2 response) and from

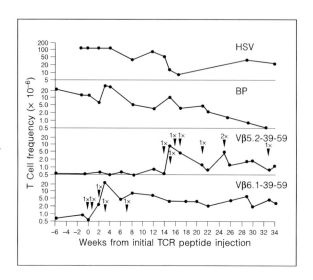

Fig. 7. Serial T-cell frequencies from patient N. L. for HSV, BP, Vβ5.2 39-59, and Vβ6.1 39-59 were determined by limiting dilution analysis as described [70]. *Arrows* = Time when the patient received peptide. Doses were as follows: 1×=100 µg; 2×=200 µg. The graph demonstrates a significant rise in T-cell frequencies to Vβ6.1 39-59 beginning 2 weeks after the first injection of peptide. Significantly elevated T-cell frequencies for Vβ6.1 39-59 persisted for over 27 weeks after the last injection with the peptide. The graph also shows a significant rise in T-cell frequencies to Vβ5.2 39-59 after the first injection of this peptide. MBP-specific T-cell frequencies were initially high and remained elevated during immunization with Vβ6.1 39-59. MBP-specific T-cell frequencies fell to low levels after immunization with Vβ5.2 39-59. The frequency of T cells specific for HSV fluctuated over the course of the study. From Bourdette et al. [70].

N. L. and other MS patients who were responsive to peptide injection [55]. These T cells were remarkably similar to rat anti-TCR peptide-reactive T cells in that they were also predominantly CD4+CD8^{dim}, expressed a variety of Vβ genes, and a combination of Th1- and Th2-like lymphokines, including IFNγ, GM-CSF, IL-4 and IL-5 (fig. 8). One notable difference was that human anti-TCR peptide-specific T cells were either MHC I or MHC II restricted. This difference may be related to the much stronger and prolonged expression of MHC II by activated human T cells than by activated rat T cells. This observation is consistent with the idea that the generation of TCR peptide-specific T cells in vivo may stem from the internal processing and association of unexpressed TCR chains with MHC molecules on the T-cell surface [69]. Only about 10% of the TCR α- and β-chains that are synthesized within the T cell are assembled and expressed on the cell surface after association with the γ, δ,

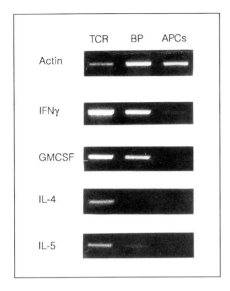

Fig. 8. Profile of lymphokine mRNA expression after antigen stimulation by Hu-BP and Vβ5.2 39-59 peptide specific T-cell lines from patient M. R., as analyzed by RT-PCR. Note the increased message production for IFNγ and GM-CSF by both lines, and for IL-4 and IL-5 by the Vβ5.2 39-59 peptide specific line, compared to APC alone. From Chon et al. [55].

and ε CD3 chains and the ζζ homodimer or ζη heterodimer (the ζ-chain is rate limiting) [75]. The remainder of the TCR α- and β-chains are degraded mainly in the lysosomal compartment, which favors MHC II loading, or in the endoplasmic reticulum, which favors MHC I loading [76]. MHC I is constitutively expressed by both rat and human T cells, providing a relatively stable source of signal for MHC I restricted TCR peptide-specific T cells. On the other hand, MHC II is expressed during activation and expansion of human T-cell clones over a period of weeks, whereas it is only briefly expressed for approximately 48 h on activated rat T cells [24]. Thus, one might expect a predominantly MHC I restricted response in rats and a mixed MHC I and II restricted response in humans. It is noteworthy that two other reports have identified MHC II restricted T-cell lines in LEW rats that were not protective against EAE [66, 67]. The situation becomes even more complex in the mouse, in which the T cells do not express MHC class II at all. Unexpectedly, mouse anti-TCR peptide specific T cells described by others were MHC class II⁻ restricted, but apparently depend on class I⁻ restricted effector cells to regulate the pathogenic T cells that only express MHC I [77, 78]. Taken together, these findings raise important questions concerning the biological functions, including protective activity and lymphokine profiles, of class I vs. class II restricted clones from both rats and humans. Hopefully, the protective function of human TCR peptide-reactive T cells can eventually be tested in the SCID-Hu model described above for testing encephalitogenic activity of human BP-reactive T cells.

Working Hypothesis

The most difficult theoretical question that remains to be addressed is why TCR peptide-specific T cells are not naturally deleted in the thymus, where there is an abundance of TCR molecules that would be expected to cause deletion of TCR-reactive cells. Recent experiments [unpubl.] have demonstrated that injection of soluble TCR peptides neonatally causes total unresponsiveness to the same peptide injected later into adult rats.

Thus, it is clear that neonatal exposure to soluble TCR peptide induces classic tolerance. Our current thinking is that MHC-associated TCR idiotopes, expressed by double-positive thymocytes rather than 'professional' thymic APC, favor positive selection of TCR-peptide-specific T cells. These TCR-specific T cells can then be expanded further as a consequence of the focused response to BP, possibly contributing to natural recovery from EAE in rats and to incomplete remissions in MS. Once these cells reach a critical threshold frequency, they can then be triggered (boosted) in vivo with small doses of peptide presented by resident APC. The TCR-specific T cells and antibodies can then home to the site of sensitization (lymph nodes) or expansion of encephalitogenic T cells (CNS) where regulation is likely to occur. The mechanism by which TCR-specific T cells regulate BP-reactive T cells is the current focus of our laboratories. Recent unpublished data suggest that soluble factors released by rat TCR-specific T cells can inhibit proliferation and passive transfer of EAE. The implication of a soluble factor raises the specter of 'bystander suppression', in which T cells specific for one TCR sequence could inhibit closely approximated encephalitogenic T cells bearing the same or different V genes.

Conclusions and Future Challenges

In this review, we have emphasized the considerable similarities between rodent and human pathogenic and regulatory T cells. Clearly, there is immunological recognition of human BP in MS patients that in some instances seems sufficient to account for the clinical and histological findings. Other considerations, notably the very low cellularity in CSF, suggest a much slower encephalitogenic process in MS than in EAE, where sensitization to BP is optimized. It is also very clear that the anti-idiotypic regulatory mechanism directed at TCR peptide sequences that we initially characterized in the rat exists in a nearly identical form in humans. If this is indeed the case, we would expect that TCR peptides will be useful as a mode of specific therapeutic intervention in human autoimmune diseases characterized by biased V gene expression. The major challenges for such an application are the timely and correct identification of V genes that are overexpressed by pathogenic T cells, and an unequivocal demonstration in humans that regulation can be mediated by TCR-specific T cells or antibodies. Assuming that these considerable challenges can be met, one can envision the rapid screening of potentially pathogenic T cells to identify one or several biased V genes, and the subsequent treatment of the patients with a personalized cocktail of preselected antigenic peptides corresponding to the relevant V genes.

Acknowledgements

The authors wish to acknowledge the scientific contributions of our Portland collaborators, Drs. Dennis N. Bourdette, Abigail C. Buenafe, Bozena Celnik, Yuan K. Chou, Richard E. Jones, Margarita R. Vainiene, Andrew D. Weinberg, Ruth H. Whitham, and the secretarial assistance of Ms. Eva Jarvie.

References

1 Alvord EC Jr: Species-restricted encephalitogenic determinants; in Alvord EC Jr, Kies MW, Suckling AJ (eds): Experimental Allergic Encephalomyelitis: A Useful Model for Multiple Sclerosis. New York, Liss, 1984, vol 146, pp 523–537.
2 Tuohy VK, Lu Z, Sobel RA, Laursen RA, Lees MB: Identification of an encephalitogenic determinant of myelin proteolipid protein for SJL mice. J Immunol 1989;142:1523–1527.
3 Kuchroo VK, Sobel RA, Yamamura T, Greenfield E, Dorf ME, Lees MB: Induction of experimental allergic encephalomyelitis by myelin proteolipid protein specific T cell clones and synthetic peptides. Pathobiology 1991;59:305–312.
4 Whitham RH, Bourdette DN, Hashim GA, Herndon RM, Ilg RC, Vandenbark AA, Offner H: Lymphocytes from SJL/J mice immunized with spinal cord respond selectively to a peptide of proteolipid protein and transfer relapsing demyelinating experimental autoimmune encephalomyelitis. J Immunol 1991;146:101–107.
5 Jones RE, Bourdette DN, Whitham RH, Offner H, Vandenbark AA: Induction of experimental autoimmune encephalomyelitis in severe combined immunodeficient mice reconstituted with allogenic or xenogeneic hematopoietic cells. J Immunol 1993;150:4620–4629.
6 Jones RE, Bourdette DN, Offner H, Vandenbark AA: Myelin basic protein-specific T cells induce demyelinating experimental autoimmune encephalomyelitis in Buffalo rats. J Neuroimmunol 1990;30:61–69.
7 Raine CS, Scheinberg LC: On the immunopathology of plaque development and repair in multiple sclerosis. J Neuroimmunol 1988;20:189–194.
8 Miller SD, Clatch RJ, Pevear DC, Trotter JL, Lipton HL: Class II-restricted T cell responses in Theiler's murine encephalomyelitis virus (TMEV)-induced demyelinating disease. I. Cross-specificity among TMEV substrains and related picornaviruses, but not myelin proteins. J Immunol 1987;138:3776–3784.
9 Miller SD, Gerety SJ, Kennedy MK, Peterson JD, Trotter JL, Tuohy VK, Waltenbaugh C, Dal-Canto MC, Lipton HL: Class II-restricted T cell responses in Theiler's murine encephalomyelitis virus (TMEV)-induced demyelinating disease. III. Failure of neuroantigen-specific immune tolerance to affect the clinical course of demyelination. J Neuroimmunol 1990;26:9–23.
10 Lehmann PV, Forsthuber T, Miller A, Sercarz EE: Spreading of T-cell autoimmunity to cryptic determinants of an autoantigen. Nature 1992;358:155–157.

11 Vandenbark AA, Vainiene M, Celnik B, Hashim G, Offner H: TCR peptide therapy decreases the frequency of encephalitogenic T cells in the periphery and the central nervous system. J Neuroimmunol 1992;39:251–260.

12 Mor F, Cohen IR: T cells in the lesion of experimental autoimmune encephalomyelitis. J Clin Invest 1992;90:2447–2455.

13 Chou YK, Bourdette DN, Offner H, Whitham R, Wang R, Hashim GA, Vandenbark AA: Frequency of T cells specific for myelin basic protein and myelin proteolipid protein in blood and cerebrospinal fluid in multiple sclerosis. J Neuroimmunol 1992;38:105–114.

14 Vandenbark AA, Bourdette DN, Whitham R, Hashim GA, Chou YK, Offner H: Episodic changes in T cell frequencies to myelin basic protein in patients with multiple sclerosis. Neurology 1993;43:2416–2417.

15 Willoughby EW, Grochowski E, Li KB, Oger J, Kastrukoff LF, Paty DW: Serial magnetic resonance scanning in multiple sclerosis: A second prospective study in relapsing patients. Ann Neurol 1989;25:43–49.

16 Buenafe AC, Vainiene M, Celnik B, Vandenbark AA, Offner H: Analysis of Vβ8-CDR3 sequences derived from the CNS of Lewis rats with experimental autoimmune encephalomyelitis. J Immunol 1994;153:386–394.

17 Offner H, Buenafe AC, Vainiene M, Celnik B, Weinberg AD, Gold DP, Hashim G, Vandenbark AA: Where, when, and how to detect biased expression of disease-relevant Vβ genes in rats with experimental autoimmune encephalomyelitis. J Immunol 1993;151:506–517.

18 Satyanarayana K, Chou YK, Bourdette D, Whitham R, Hashim GA, Offner H, Vandenbark AA: Epitope specificity and V gene expression of cerebrospinal fluid T cells specific for intact versus cryptic epitopes of myelin basic protein. J Neuroimmunol 1993;44:57–68.

19 Offner H, Hashim GA, Chou YK, Celnik B, Jones R, Vandenbark AA: Encephalitogenic T cell clones with variant receptor specificity. J Immunol 1988;141:3828–3832.

20 Whitaker JN, Gupta M, Smith OF: Epitopes of immunoreactive myelin basic protein in human cerebrospinal fluid. Ann Neurol 1986;20:329–336.

21 Weinberg AD, Wallin JJ, Jones RE, Sullivan TJ, Bourdette DN, Vandenbark AA, Offner H: Target organ specific upregulation of the MRC OX-40 marker and selective production of Th1 lymphokine mRNA by encephalitogenic T helper cells isolated from the spinal cord of rats with experimental autoimmune encephalomyelitis. J Immunol 1994;152:4712–4721.

22 Vandenbark AA, Hashim G, Offner H: Myelin basic protein, MHC restriction molecules and T cell repertoire. Prog Clin Biol Res 1990;336:93–108.

23 Vandenbark AA, Offner H, Reshef T, Fritz R, Chou CHJ, Cohen IR: Specificity of T lymphocyte lines for peptides of myelin basic protein. J Immunol 1985B;135:229–233.

24 Offner H, Brostoff SW, Vandenbark AA: Antibodies against I-A and I-E determinants inhibit the in-vitro activation of an encephalitogenic T lymphocyte line. Cell Immunol 1986;100:364–373.

25 Burns FR, Li X, Shen N, Offner H, Chou Y, Vandenbark AA, Heber-Katz E: Both rat and mouse T cell receptors specific for the encephalitogenic determinant of myelin basic protein use similar Vα and Vβ chain genes even though the major histocompatibility complex and encephalitogenic determinants being recognized are different. J Exp Med 1989;169:27–39.

26 Gold DP, Offner H, Sun D, Wilery S, Vandenbark AA, Wilson DB: Analysis of T cell receptor β chains in Lewis rats with experimental allergic encephalomyelitis: Conserved complementary determining region 3. J Exp Med 1991;174:1467–1476.

27 Zhang X-M, Heber-Katz E: T cell receptor sequences from encephalitogenic T cells in adult Lewis rats suggest an early ontogenic origin. J Immunol 1992;148:746–752.

28 Offner H, Hashim GA, Celnik B, Galang A, Li X, Burns FR, Shen N, Heber-Katz E, Vandenbark AA: T cell determinants of myelin basic protein include a unique encephalitogenic I-E-restricted epitope for Lewis rats. J Exp Med 1989;170:355–367.

29 Vandenbark AA, Hashim GA, Celnik B, Galang A, Li X, Heber-Katz E, Offner H: Determinants of human myelin basic protein that induce encephalitogenic T cells in Lewis rats. J Immunol 1989;143:3512–3516.

30   Hashim GA, Galang AB, Srinivasen JV, Carvalho EF, Offner H, Vandenbark AA, Cleveland WL, Day ED: Defective T helper cell epitope responsible for the failure of region 69–84 of the human MBP to induce EAE in the Lewis rat. J Neurosci Res 1989;24:222–230.

31   Vandenbark AA, Vainiene M, Cenik B, Hashim GA, Buenafe A, Offner H: Definition of encephalitogenic and immunodominant epitopes of guinea pig myelin basic protein (Gp-BP) in Lewis rats tolerized neonatally with Gp-BP or Gp-BP peptides. J Immunol 1994;in press.

32   Offner H, Vainiene M, Gold DP, Celnik B, Wang R, Hashim GA, Vandenbark A: Characterization of the immune response to a secondary encephalitogenic epitope of basic protein in Lewis rats. I. T cell receptor peptide regulation of T cell clones expressing cross-reactive Vβ genes. J Immunol 1992;148:1706–1711.

33   Gold DP, Vainiene M, Celnik B, Wiley S, Gibbs C, Hashim GA, Vandenbark AA, Offner H: Characterization of the immune response to a secondary encephalitogenic epitope of basic protein in Lewis rats. II. Biased T cell receptor Vβ expression predominates in spinal cord infiltrating T cells. J Immunol 1992;148:1712–1717.

34   Bourdette DN, Vainiene M, Morrison WJ, Jones R, Turner J, Hashim GA, Vandenbark AA, Offner H: Myelin basic protein specific T cell lines and clones derived from the central nervous system of rats with experimental autoimmune encephalomyelitis only recognize encephalitogenic epitopes. J Neurosci Res 1991;30:308–315.

35   Vainiene M, Offner H, Morrison WJ, Wilkenson M, Vandenbark AA: Clonal diversity of basic protein specific T cells in Lewis rats recovered from experimental autoimmune encephalomyelitis. J Neuroimmunol 1991;33:207–216.

36   Offner H, Vainiene M, Gold DP, Morrison WJ, Wang RY, Hashim GA, Vandenbark AA: Protection against experimental encephalomyelitis: Idiotypic autoregulation induced by a non-encephalitogenic T cell clone expressing a cross-reactive T cell receptor V gene. J Immunol 1991B;146:4165–4172.

37   Vandenbark AA, Chou YK, Bourdette D, Whitham R, Chilgren J, Chou C-HJ, Konat G, Hashim G, Vainiene M, Offner H: Human T lymphocyte response to myelin basic protein: Selection of T lymphocyte lines from MBP-responsive donors. J Neurosci Res 1989;23:21–30.

38   Chou YK, Vainiene M, Whitham R, Bourdette D, Chou C-HJ, Hashim G, Offner H, Vandenbark AA: Response of human T lymphocyte lines to myelin basic protein: Association of dominant epitopes with HLA class II restriction molecules. J Neurosci Res 1989;23:207–216.

39   Chou YK, Henderikx P, Vainiene M, Whitham R, Bourdette D, Chou C-HJ, Hashim GA, Vandenbark AA: Specificity of human T cell clones reactive to immunodominant epitopes of myelin basic protein. J Neurosci Res 1991;28:280–290.

40   Chou YK, Henderikx P, Jones RE, Kotzin B, Hashim GA, Offner H, Vandenbark AA: Human CD8+ T cell clone regulates autologous CD4+ myelin basic protein specific T cells. Autoimmunity 1992;14:111–110.

41   Offner H, Hashim G, Chou YK, Bourdette D, Vandenbark AA: Prevention, suppression and treatment of EAE with a synthetic T cell receptor V region peptide; in Alt FW, Vogel MG (eds): Molecular Mechanisms of Immunological Self-Recognition. San Diego, Academic Press, 1993, pp 199–230.

42   Heber-Katz E, Acha-Orbea H: The V-region disease hypothesis: Evidence from autoimmune encephalomyelitis. Immunol Today 1989;10:164–169.

43   Goverman J, Woods A, Larson L, Weinver LP, Hood L, Zaller DM: Transgenic mice that express a myelin basic protein-specific T cell receptor develop spontaneous autoimmunity. Cell 1993;72:551–560.

44   Whitham RH, Kotzin BL, Buenafe AC, Weinberg AD, Jones RE, Hashim GA, Hoy CM, Vandenbark AA, Offner H: Treatment of relapsing experimental autoimmune encephalomyelitis with T cell receptor peptides. J Neurosci Res 1993;35:115–128.

45   Davis MM, Bjorkman PJ: T-cell antigen receptor genes and T-cell recognition. Nature 1988; 334:395–402.

46   Williams WV, Weiner DB, Wadsworth S, Greene MI: The antigen-major histocompatibility complex-T cell receptor interaction: A structural analysis. Immunol Rev 1988;7:339–344.

47   Claverie JM, Prochnicka-Chalufour A, Bougueleret A: Implications of a Fab-like structure for the T-cell receptor. Immunol Today 1989;10:10–14.

48   Marrack P, Kappler J: The staphylococcal enterotoxins and their relatives. Science 1990;248: 705–711.

49   Wucherpfennig DW, Ota K, Endo N, Seidman JG, Rosenzweig A, Weiner HS, Hafler DA: Shared human T cell receptor Vβ usage to immunodominant regions of myelin basic protein. Science 1990;248:1016–1019.

50   Ben-Nun A, Liblau RS, Cohen L, Lehmann D, Tournier-Lasserve E, Rosenzweig A, Jingwu Z, Raus JCM, Bach MA: Restricted T-cell receptor Vβ gene usage by myelin basic protein-specific T cell clones in multiple sclerosis: Predominant genes vary in individuals. Proc Natl Acad Sci USA 1991;88:2466–2470.

51   Kotzin BL, Karuturi S, Chou YK, Lafferty J, Forrester JM, Better M, Nedwin GE, Offner H, Vandenbark AA: Preferential T cell receptor Vβ gene usage in myelin basic protein reactive T cell clones from patients with multiple sclerosis. Proc Natl Acad Sci USA 1991;88:9161–9165.

52   Martin R, Utz U, Coligan JE, Richert JR, Flerlage M, Robinson E, Stone R, Biddison WE, McFarlin DE, McFarland HF: Diversity in fine specificity and T cell receptor usage of the human CD4+ cytotoxic T cell response specific for the immunodominant myelin basic protein peptide 87–106. J Immunol 1992;148:1359–1366.

53   Oksenberg JR, Panzara MA, Begovich AB, Mitchell D, Erlich HA, Murray RS, Shimonkevitz R, Sherritt M, Rothbard J, Bernard CCA, Steinman L: Selection of T-cell receptor Vβ-Dβ-Jβ gene rearrangements with specificity for a myelin basic protein peptide in brain lesions of multiple sclerosis. Nature 1993;362:68–70.

54   Chou YK, Buenafe AC, Dedrick R, Morrison WJ, Bourdette DN, Whitham R, Atherton J, Lane J, Spoor E, Hashim GA, Offner H, Vandenbark AA: T cell receptor Vβ gene usage in the recognition of myelin basic protein by cerebrospinal fluid- and blood-derived T cells from patients with multiple sclerosis. J Neurosci Res 1994;37:169–181.

55   Chou YK, Morrison WJ, Weinberg AD, Dedrick R, Whitham R, Bourdette DN, Hashim G, Offner H, Vandenbark AA: Immunity to T cell receptor peptides in multiple sclerosis. II. T cell recognition of Vβ5.2 and Vβ6.1 CDR2 peptides. J Immunol 1994;152:2520–2529.

56   Chou YK, Jones RE, Bourdette D, Whitham R, Hashim G, Atherton J, Offner H, Vandenbark AA: Human myelin basic protein (MBP) epitopes recognized by mouse MBP-selected T cell lines from multiple sclerosis patients. J Neuroimmunol 1994;49:45–50.

57   Saeki Y, Mima T, Sakoda S, Fujimura H, Arita N, Nomura T, Kishimoto T: Transfer of multiple sclerosis into severe combined immunodeficiency mice by mononuclear cells from cerebrospinal fluid of the patients. Proc Natl Acad Sci USA 1992;89:6157–6161.

58   Margalit H, Spouge JL, Cornette JL, Ease KB, DeLisi C, Bersofsky JA: Prediction of immunodominant helper T cell antigenic sites from the primary sequence. J Immunol 1987;138: 2213–2229.

59   Rothbard JB, Taylor WR: A sequence pattern common to T cell epitopes. EMBO J 1988;7: 93–100.

60   Vandenbark AA, Hashim G, Offner H: Immunization with a synthetic T-cell receptor V-region peptide protects against experimental autoimmune encephalomyelitis. Nature 1989;341: 541–544.

61   Hashim GA, Vandenbark AA, Galang AB, Diamanduros T, Carvalho E, Srinivasan J, Jones R, Vainiene M, Morrison WJ, Offner H: Antibodies specific for a Vβ8 T cell receptor peptide suppress experimental autoimmune encephalomyelitis. J Immunol 1990;144:4621–4627.

62   Stevens DB, Karpus WJ, Gould KE, Swanborg RH: Studies of Vβ8 T cell receptor peptide treatment in experimental autoimmune encephalomyelitis. J Neuroimmunol 1992;37:123–129.

63   Howell MD, Winters ST, Olee T, Powell HC, Carlo DJ, Brostoff SW: Vaccination against experimental allergic encephalomyelitis with T cell receptor peptides. Science 1989;246:668–670.

64   Desquenne-Clark L, Esch TR, Otvos L Jr, Heber-Katz E: T cell receptor peptide immunization leads to enhanced and chronic experimental allergic encephalomyelitis. Proc Natl Acad Sci USA 1991;88:7219–7223.

65   Kawano YI, Sasamoto Y, Kotake S, Thurau SR, Wiggert B, Gery I: Trials of vaccination against experimental autoimmune uveoretinitis with a T-cell receptor peptide. Curr Eye Res 1991;10: 789–795.

66   Sun D: Synthetic peptides representing sequence 39 to 59 of rat Vβ8 TCR fail to elicit regulatory T cells reactive with Vβ8 TCR on rat encephalitogenic T cells. Cell Immunol 1992;141: 200–210.

67   Jung S, Schluesener HJ, Toyka KV, Hartung H-P: Modulation of EAE by vaccination with T cell receptor peptides: Vβ8 T cell receptor peptide-specific CD4 lymphocytes lack direct immunoregulatory activity. J Neuroimmunol 1993;45:15–22.

68   Offner H, Hashim GA, Vandenbark AA: T cell receptor peptide therapy triggers autoregulation of experimental encephalomyelitis. Science 1991;251:430–432.

69   Vandenbark AA, Hashim G, Offner H: TCR peptide therapy in autoimmune disease; in Sercarz E (ed): Regulation of Autoimmunity: EAE/MS. Int Rev Immunol New York, Harwood Academic Publishers, 1993, vol 9, pp 251–276.

70   Bourdette DN, Whitham RH, Chou YK, Morrison WJ, Atherton J, Kenny C, Liefeld D, Hashim GA, Offner H, Vandenbark AA: Immunity to T cell receptor peptides in multiple sclerosis. I. Successful immunization of patients with synthetic Vβ5.2 and Vβ6.1 CDR2 peptides. J Immunol 1994;152:2510–2519.

71   Vainiene M, Gold DP, Celnik B, Hashim GA, Vandenbark AA, Offner H: Common sequence on distinct Vβ genes defines a protective idiotope in experimental encephalomyelitis. J Neurosci Res 1992;31:413–420.

72   Weinberg AD, Celnik B, Vainiene M, Buenafe AC, Vandenbark AA, Offner H: The effect of TCR Vβ8 peptide therapy on T cell populations isolated from the spinal cords of Lewis rats with experimental autoimmune encephalomyelitis. J Neuroimmunol 1994;49:161–170.

73   Weinberg AD, Wyrick G, Celnik B, Vainiene M, Bakke A, Offner H, Vandenbark AA: Lymphokine mRNA expression in the spinal cords of Lewis rats with EAE is mainly associated with a host recruited CD45R hi/CD4+ population during recovery. J Neuroimmunol 1993;48: 105–118.

74   Hashim GA, Offner H, Wang RY, Shukla K, Carvalho E, Morrison WJ, Vandenbark AA: Spontaneous development of protective anti-T cell receptor autoimmunity targeted against a natural EAE-regulatory idiotope located within the 39–59 region of the TCR-Vβ8.2 chain. J Immunol 1992;149:2803–2809.

75   Klausner RD, Lippincott-Schwartz J, Bonifacino JS: The T cell antigen receptor: Insights into organelle biology. Ann Rev Cell Biol 1990;6:403–431.

76   Minami Y, Weissman AM, Samelson LE, Klausner RD: Building a multichain receptor: Synthesis, degradation and assembly of the T-cell antigen receptor. Proc Natl Acad Sci USA 1987; 2688–2692.

77   Kumar V, Sercarz EE: The involvement of T cell receptor peptide-specific regulatory CD4+ T cells in recovery from antigen-induced autoimmune disease. J Exp Med 1993;178:909–916.

78   Gaur A, Haspel R, Mayer JP, Fathman CG: Requirement for CD8+ T cells in T cell receptor peptide-induced clonal unresponsiveness. Science 1993;259:91–94.

Arthur A. Vandenbark, PhD, Neuroimmunology Research 151D,
V. A. Medical Center, Portland, OR 97201 (USA)

Adorini L (ed): Selective Immunosuppression: Basic Concepts and Clinical Applications.
Chem Immunol. Basel, Karger, 1995, vol 60, pp 126–149

....

Antigen-Specific Immunosuppression: Oral Tolerance for the Treatment of Autoimmune Disease

David A. Hafler [a], *Howard L. Weiner* [b]

[a] Laboratory of Molecular Immunology and
[b] The MS Clinical and Research Unit, Center for Neurologic Diseases, Brigham and
Women's Hospital, and Harvard Medical School, Boston, Mass., USA

The Etiology of Multiple Sclerosis

Multiple sclerosis (MS) is a chronic inflammatory disease of central nervous system myelin characterized by focal T cell and macrophage infiltrates that lead to demyelination [1–3] and loss of neurologic function. In acute MS plaques, activated T cells secreting regulatory cytokines and expressing growth factor receptors for IL-2 as well as activated, class II MHC positive macrophages are present [4–6]. This active inflammatory process is confined to the white matter in the central nervous system, not affecting either the peripheral nervous system or other organs. Although it is generally accepted that this CNS inflammatory process causes demyelination and the resulting neurologic disability in MS, the mechanism(s) by which the inflammation is initiated and maintained both in this disease, and in fact in all of the presumed T-cell-mediated autoimmune diseases in humans has been difficult to define.

There are two primary explanations for the CNS inflammation in MS. Either the brain and spinal cord are infected by a virus or other infectious agent and cells that infiltrate the CNS are targeted to the infectious agent, or the infiltrating cells are autoimmune in nature and attack normal myelin proteins. At our present state of knowledge, either hypothesis is possible. While the inability to identify an infectious agent in the CNS of patients with MS to date does not prove that such an agent is not present, recent experiments have provided new information regarding autoreactive T-cell responses to the CNS. This has led to the postulate that viruses and infectious agents, perhaps acting as superantigens, trigger or drive the autoimmune process, as opposed to being the primary CNS target of infiltrating cells. This review will first present evidence for an autoimmune mechanism in MS, followed by a discussion of oral tolerance as a method to treat the disease.

Myelin Antigens as Targets for T Cells

What are the characteristics of a tissue specific autoantigen that will be recognized by T cells? Approximately 0.1% of a restricting class II MHC protein must be occupied with the antigen peptide for T-cell activation to occur [7]. Considering there is competition by a large variety of self and foreign peptide for binding to MHC class II proteins [8], it is possible that this requirement may be met only by peptides which are present in relatively large quantities in antigen presenting cells, processed efficiently and bound with a high affinity to class II molecules. For example, myelin basic protein (MBP) as a cationic protein may be efficiently processed by antigen-presenting cells since cationization of antigens has been shown to increase their uptake and processing by macrophages. Thus, the physical characteristics and processing of the antigen in the tissue may limit the number of potential autoantigens that the immune system is capable of recognizing.

Secondly, a T-cell autoantigen must be processed in such a way as to be presented by class II MHC by the antigen-presenting cells specific for that tissue site (fig. 1). CNS microglia, which are specialized CNS macrophages, might phagocytose breakdown products of myelin and present peptide antigens to T cells [4, 9]. Alternatively, astrocytes, which after activation can express class II MHC, might present CNS antigens in the context of class II MHC [10, 11]. Oligodendrocytes, which synthesize the myelin membrane, do not express class II MHC and thus can not present antigen to T cells [12]. Lastly, it is possible that myelin antigens may also be presented by cerebral endothelial cells [13]. While it is not known what the antigen-presenting cell is in the CNS, experiments in the EAE model clearly indicate that at least two antigens, MBP and PLP, can be recognized by activated autoreactive T cells in the CNS and cause severe CNS damage associated with neurologic defects. In contrast, activated myelin oligodendrocyte glycoprotein (MOG) and S-100 reactive T-cell clones can also home to the CNS and recognize their cognate antigen with associated pathologic inflammation but with only minor if any neurologic defects.

Autoreactive T Cells in MS

MBP and PLP are the two most abundant myelin proteins and are primary candidate autoantigens in MS because of their ability to induce EAE. As mentioned above, other potentially important myelin antigens included MOG

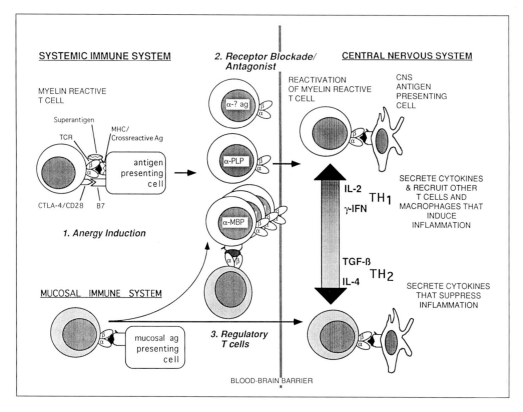

SYSTEMIC IMMUNE SYSTEM   2. Receptor Blockade/   CENTRAL NERVOUS SYSTEM
                             Antagonist

MYELIN REACTIVE   REACTIVATION   CNS
T CELL   OF MYELIN REACTIVE ANTIGEN
                                             T CELL   PRESENTING
                                                                 CELL
      Superantigen   α-? ag
              MHC/
    TCR   Crossreactive Ag
                     antigen   α-PLP
                     presenting
                        cell
CTLA-4/CD28   B7   SECRETE CYTOKINES
                                             IL-2   & RECRUIT OTHER
                           α-MBP   TH1   T CELLS AND
    1. Anergy Induction   γ-IFN   MACROPHAGES THAT
                                                                 INDUCE
                                                                 INFLAMMATION

                                             TGF-ß
MUCOSAL IMMUNE SYSTEM   TH2
                                             IL-4   SECRETE CYTOKINES
                                                                 THAT SUPPRESS
                                                                 INFLAMMATION

                     mucosal ag   3. Regulatory
                     presenting   T cells
                        cell

                    BLOOD-BRAIN BARRIER

Fig. 1. Overview of proposed pathophysiology for MS and specific areas where the immune system can be manipulated. The minimal requirements in mammals for inducing inflammatory autoimmune disease in the CNS white matter are activated CD4+ T cells recognizing MBP or PLP. In MS, we hypothesize that some activating event, such as a viral superantigen or molecular mimicry activates an autoreactive T cell. This allows the activated T cell to migrate into the CNS where it can now recognize CNS antigen in the context of MHC, recruiting effector cells leading to the inflammatory CNS response. The process may be blocked by (1) anergy induction, (2) regulatory T cells, or (3) receptor blockade or antagonist (see text).

and myelin associated glycoprotein (MAG), though neither protein can induce clinical EAE in laboratory animals. Autoantibodies against MBP and other myelin components have been demonstrated in MS [14–16], but they are not believed to play a primary role in disease pathogenesis given the prominent cellular infiltrate in the CNS.

There have been many investigations of MBP and PLP reactivity in MS using 7-day proliferation assays of whole peripheral blood mononuclear cells

[17, 18]. These investigations have shown a slight increase in T-cell responses to human MBP in subjects with MS as compared to normal subjects or other neurologic disease patients, but the magnitude of the difference has generally been less then convincing. T-cell cloning techniques have been applied to study autoreactive T cells both in the peripheral blood and spinal fluid. Antigen-specific T-cell clones are generated by repeated stimulation and culture of T cells with antigen in the presence of T-cell growth factors.

In the EAE model, T-cell clones can be derived from animals injected with MBP in adjuvant that recognize immunodominant regions of MBP that are influenced by the MHC of the animal. These clones can transfer EAE when injected into naive animals [19–22]. For example in Lewis rats, the MBP (68–88) peptide is immunodominant and is the epitope for the majority of encephalitogenic T cell clones [20–22]. Lewis rat T-cell clones generated against MBP (residues 43–67) are only weakly able to cause disease.

In humans, immunodominant regions of MBP that are presented by defined MHC proteins have been defined. Immunodominant regions of MBP between residues 84–106 and 143–172 near the C-terminus [23–27; see also 57, 60, 61, 62 and review in 58]. In DR2+ individuals, a high proportion of T-cell clones reactive with MBP recognized the 84–106 region, whether or not they had MS [23]. Both DR and DQ antigens were found to be restricting elements indicating that immunodominant MBP peptides can be presented by different class II molecules. It has also been demonstrated that T cells from genetically diverse individuals responding to different epitopes of human MBP are associated with distinct MHC class II molecules [28]. Of interest was the finding that DR2 possessed an unusual capacity to restrict all of the epitopes identified on MBP [25, 26, 28]. Moreover, different types of DR2 have the capacity to present a number of different MBP epitopes [29]. These data together indicate that disease-associated DR2 antigens can present a variety of MBP peptides to T cells and this is likely to be associated with competition for peptide binding between different MBP regions. Thus, peptide competition for binding to class II molecules may in part determine immunodominant sites of an autoantigen.

Another approach in defining autoreactive T cells in MS has been to use a single cell 'immunospot' assay in which T cells reactive with a particular antigen secrete gamma-interferon. A series of studies using this approach has demonstrated autoreactive T cells in the blood and CSF of MS patients that react with MBP, PLP, and MOG. B-cell responses to MBP were also reported to be increased [30–32]. These data indicate that measures such as cytokine secretion can be used to define antigen-reactive clones in the blood of MS patients. Whether there will be a specific pattern of cytokines secreted by autoreactive T-cell clones remains to be defined.

T-Cell Receptor Characterization in MS

There appears to be a moderately restricted TCR usage among human T-cell clones specific for immunodominant regions of MBP. Among reports from different laboratories, common TCR usage in recognition of immunodomiant MBP peptides were found among T-cell clones isolated from single individuals [33–35]. In contrast, a broader repertoire of TCRs can be found among different individuals recognizing the same MBP peptide, though again, the TCR usage does not appear to be totally random [33–36]. More recently, we have gone on to sequence the TCR used in a large panel of MBP reactive T-cell clones [37]. The majority of MBP-specific T-cell lines recognized the immunodominant MBP(84–102) and MBP(143–168) peptides and were HLA-DR restricted. Though T cells reactive with these epitopes expressed different TCR Vα & Vβ chains, Vα3.1 and Vα8 were more frequently observed while Vβ usage was more diverse. In 2 patients with the DR2 haplotype, the response to MBP was focused on the MBP(84–102) peptide. In both patients, in vivo expanded population(s) (three expanded populations in the first patient, one expanded population in the second) dominated the response. Two MBP(84–102)-specific T-cell clones from a normal subject with the DR2 haplotype also had identical TCR sequences. Clonality was proven by demonstrating identical TCR α- and β-chain sequences and identical sequences of a TCR γ-chain or of a second TCR α-chain rearrangement in independent clones. Repeated analysis of a patient after 13 months demonstrated that expanded clones had persisted in vivo. A representative of one of the expanded clones was again obtained after 31 months by IL-2 stimulation suggesting this clone was activated in vivo. These data indicate that while the recognition of human MBP can involve the use of different TCR V genes, the immune response is dominated in some DR2 individuals by expanded clones that may persist in vivo for relatively long periods of time.

Assuming that T cells recognizing MBP peptides caused MS, it is clear that while both the number of immunodominant peptides and T-cell receptors used to recognize the peptides are not random, the repertoire is broad enough to make such an approach highly problematic. Instead, antigen-specific therapies not requiring knowledge of a particular peptide/TCR dominant for an individual would be preferable.

Activation of Autoreactive T Cells

T cells recognizing MBP and PLP exist in normal humans. From investigation of the EAE model, it has been learned that for these cells to be pathogenic they must be activated in vivo. To examine whether MBP reactive T cells are activated in vivo, an hprt⁻ mutant assay was used [38]. The assay is based on the observation that dividing cells acquire random mutations during DNA synthesis. Some of these mutations occur in the hprt gene which results in inactivation of the hprt enzyme. Thus, mutant cells do not metabolize thioguanine to a cytotoxic metabolite, which allows a very effective selection of these mutants in culture. Eleven of 258 mutant T cell clones cultured by mitogen from the peripheral blood of 5 of 6 MS patients showed strong reactivity to MBP, while none of 114 clones grown from blood of normal subjects did. However, clones were not investigated from subjects with other neurologic diseases or with nonmyelin antigens. These data suggest that MBP-reactive T cells are activated in MS patients and thus are pathogenic.

We directly investigated in a total of 72 subjects with definitive MS as to whether myelin-reactive T cells, which might be critical for the pathogenesis of MS, exist in a different state of activation as compared to myelin reactive T cells cloned from the blood of normal individuals. While there were no differences in the frequencies of MBP- and PLP-reactive T cells after primary antigen stimulation, the frequency of MBP or PLP but not tetanus toxoid reactive T cells generated after primar rIL-2 stimulation were significantly higher in MS patients as compared to control individuals. Primary rIL-2 stimulated MBP-reactive T-cell lines were CD4+ and recognized MBP epitopes 84–102 and 143–168 similar to MBP-reactive T-cell lines generated with primary MBP stimulation. In the CSF of MS patients, MBP-reactive T cells generated with primary rIL-2 stimulation accounted for 7% of the IL-2-responsive cells, greater than 10-fold higher than paired blood samples, and these T cells also selectively recognized MBP peptides 84–102 and 143–168. In striking contrast, MBP-reactive T cells were not detected in CSF obtained from patients with other neurologic diseases. These results provide definitive in vitro evidence of an absolute difference in the activation state of myelin reactive T cells in the central nervous system of patients with MS and provides evidence of a pathogenic role of autoreactive T cells in the disease.

If circulating, autoreactive T cells are present in the circulation of normal individuals, how do they become activated (fig. 1)? The mechanism is of particular interest in a CNS disease such as MS, since resting T cells appear unable to cross the blood-brain barrier. Possible mechanisms in the absence of autoantigen involve immune activation associated with infections which include: (1) molecular mimicry, (2) activation by superantigens, (3) bystander CD2 activation.

(1) *Molecular mimicry* implies that epitopes of an infectious agent such as bacteria or viruses which induce an immune response in the host are present on self proteins of the host. The immune response directed against such cross-reactive epitopes of the foreign invader may possibly result in the activation of autoreactive T cells and autoimmunity. Such a mechanism could explain the autoimmune response in postviral encephalomyelitis which occurs at a time

when the virus is cleared as a result of the antiviral immune response of the host. An example of this is the hepatitis B virus polymerase, which cross-reacts with an epitope of myelin basic protein. Injection of this hepatitis virus protein with adjuvant into rabbits induces an inflammatory CNS disease as evidenced by histological analysis [39].

(2) *Superantigens*, which bind to class II antigens, and specific TCR Vβ segments may activate T cells [40]. Such superantigen activation may occur during the course of bacterial or viral infections. As superantigens drive T-cell expansion which use specific TCRs, it is possible that the oligoclonal T cells in the blood and CSF of MS patients are driven by exposure to a superantigen. Thus, the observation of oligoclonal T cells in MS, which appears to be a common finding of T cells cloned out of the CSF and TCR mRNA isolated from brain tissue, may be a 'footprint' of superantigen activation of T cells which may not be necessarily involved in the disease process. This also raises the possibility that the oligoclonal immunoglobulin bands found in MS CSF are also related to T cell activation by superantigen.

(3) *Bystander CD2 activation.* There are two major pathways for activating a T cell; the CD3/TCR pathway, and the CD2 activation pathway [41–43]. We have recently demonstrated that T cells, activated via the CD2 pathway by adhesion molecules such as LFA 3 and activated T cells expressing high surface densities of adhesion molecules, can trigger resting T cells to proliferate via these adhesion pathways [44]. Activated B cells may also stimulate T cells. Thus, we postulate that the activation of T cells in association with a viral or other infection could in theory lead to activation of autoreactive T cells in a bystander fashion and thus trigger an autoimmune response.

Antigen-Specific Immunotherapy

Overview of Antigen-Specific Immunotherapy
In the section above, we presented an overview for an autoimmune hypothesis in the etiology of MS. In total, these experiments suggest a number of manipulations that may be used to block the immune response. It is important to note the likelihood that no in vitro experiment can prove the autoimmune hypothesis for MS or perhaps any of the presumed autoimmune diseases. Instead, only specific manipulations in vivo where antigen-reactive cells are targeted with associated amelioration of disease activity will allow the understanding of the disease's pathogenesis. Thus, these specific manipulations of the immune response both attempts to treat the disease as well as scientific experiments to understand the disease.

There are two general aproaches to antigen specific immunotherapy of autoimmune disease (fig. 1). Assuming MS is mediated by myelin reactive T cells, *one approach is to block either the initial activation or the subsequent recognition of autoantigen in the CNS of this autoreactive T cells.* This approach requires the immune response against the myelin antigen to be restricted in terms of either the number of epitopes recognized or the TCR repertoire that is used. This includes MHC blocking peptides, TCR antagonists, and TCR peptides which may inhibit specific T-cell function. Unfortunately, it would appear that although there are MBP and PLP dominant epitopes and some-what restricted use of TCRs in their recognition, the outbred human species contains too many myelin-reactive T cells with different TCR usage to success-fully inhibit the total of the spreading immune response to myelin antigens.

The *second approach involves the antigen-specific targeting of T cells that downregulate immune responses to the inflammatory sites.* In this instance, it is not necessary to know the inciting antigen that elucidates the immune response nor is it required that the T cells be restricted in terms of antigen recognition or TCR usage. This approach may well be physiologic in regulating the normal immune response. One such method which has attracted much attention re-cently is the use of oral tolerization where the autoantigen is delivered orally with the generation of T cells secreting TGF-β, IL-4, and IL-10 that migrate to the site of inflammation and suppress the immune response. Our laboratories have been involved in using a number of these two approaches to treat auto-immune disease. As we believe this second approach represents the best method now available to treat T-cell-mediated autoimmune disease, we will discuss it in greatest detail below.

Oral Tolerance

Oral tolerance represents the exogenous administration of antigen to the peripheral immune system via the gut. As such, it is a form of antigen-driven peripheral immune tolerance. Immunologic tolerance is not programmed into the germ line but is acquired during maturation of the immune system by mechanisms that delete or inactivate antigen-reactive clones. There are three basic mechanisms to explain antigen-driven tolerance: clonal deletion, clonal anergy, and active suppression [45, 46]. A large number of studies have shown that one of the primary mechanims associated with oral tolerance is the gener-ation of active suppression [47]. More recently, clonal anergy has been demon-strated [48, 49]. There is little evidence that orally administered antigen in-duces clonal deletion. During the course of our investigations we have delineated two pathways by which oral tolerization results in systemic hypo-responsiveness by either active suppression or clonal anergy [50]. These path-ways are described below in a schema which forms the theoretical basis for this review.

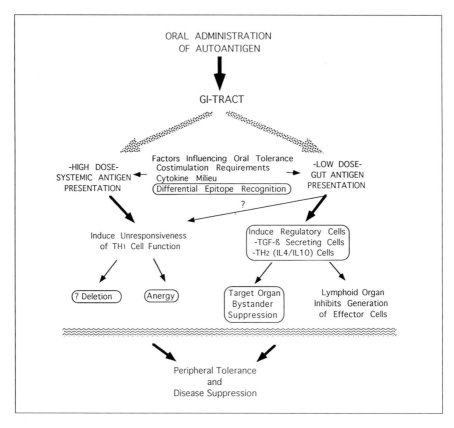

Fig. 2. Mechanisms of oral tolerance. High-dose antigen administration leads to the generation of anergy, while lower dose oral antigen induces regulatory T cells that secrete TGF-β1 and/or IL-4/IL-10 which migrate to the CNS and downregulate the local immune response.

Mechanism of Oral Tolerance (Schema)

Based on recent findings in our laboratory and others, we believe that oral tolerance can be viewed as *anergy-driven oral tolerance* or *regulatory-cell-driven oral tolerance*. The primary factor which determines which form of peripheral tolerance develops following oral administration of antigen is the dose of antigen fed. Low doses of antigen favor the generation of active suppression or regulatory cell-driven tolerance whereas high doses of antigen favor anergy-driven tolerance (fig. 2). Although these forms of oral tolerance are not mutually exclusive and may occur simultaneously, they are distinct and the use of oral tolerance to treat autoimmune diseases is critically dependent on which of these two mechanisms is triggered.

The delineation of these two mechanisms of oral tolerance was based on the following: (1) investigations in our laboratory in which low doses of orally administered autoantigens were shown to suppress experimental autoimmune diseases via the generation of regulatory cells that suppressed in vitro and in vivo via the secretion of downregulatory cytokines such as TGF-β [51, 110]; (2) investigations from other laboratories which demonstrated clonal anergy following oral administration of large doses of antigen with no evidence of active suppression [48, 49]; (3) a large series of investigations demonstrating transferrable suppression following oral tolerance [3] including work which showed two components of oral tolerance, one that was abrogated by treatment with low dose cyclophosphamide and one that was not, a difference that was dose dependent [52], and (4) direct comparison in our laboratory demonstrating that the two mechanisms depend on the dose [50] (fig. 2).

As shown in the schema, *low doses* of antigen result in the generation of antigen-specific regulatory cells and as such involve the presentation of antigen by gut-associated antigen-presenting cells. Such presentation preferentially induces regulatory cells which upon subsequent recognition of antigen in vivo or in vitro, secrete the suppressive cytokine TGF-β. In addition, Th2 responses are preferentially generated in the gut, resulting in cells which secrete IL-4 and IL-10. These antigen-specific regulatory cells migrate to lymphoid organs and suppress immune responses by inhibiting the generation of effector cells and to the target organ and suppress disease by releasing antigen nonspecific cytokines (bystander suppression). Several factors can affect the generation of regulatory cells including costimulation requirements, the cytokine milieu in which the immune response is generated and differential generation of epitopes which preferentially may trigger certain regulatory cells.

High doses of orally administered antigen result in systemic antigen presentation after antigen passes through the gut and enters the systemic circulation either as intact protein or antigen fragments. High doses of antigen induces unresponsiveness of Th1 cell function, primarily via clonal anergy. The degree to which clonal anergy following high doses of antigen merely represents the direct passage of small amounts of antigen into the systemic or portal circulation or is dependent on filtration by the gut is unknown. Why there is reduced active suppression with high doses of orally administered antigen is unclear, but could relate to anergizing cells involved in the generation of active suppression. In addition, the degree to which costimulatory requirements, cytokine milieu and differential epitope recognition may preferentially favor the generation of anergy in Th1 cells is not known.

Active Suppression

Many studies demonstrate the active suppression is an important mechanism for oral tolerance [47, 53–58]. After feeding antigens such as ovalbumin or sheep red blood cells, transferrable suppression mediated by T cells from Peyer's patches, mesenteric lymph node and spleen has been demonstrated. Investigators have also reported initial sensitization prior to the appearance of suppression [54]. Further characterization of active suppression as measured in these systems has not occurred most probably related to the difficulties in defining the biology of suppression [59, 60]. Nonetheless, the demonstration of transferable cellular suppression associated with oral tolerance is a recurrent theme reported by many investigators [47]. Of note are the studies of Mowat et al. [52] who reported that high doses of ovalbumin induced tolerance not abrogated by cyclophosphamide and such tolerance-affected antibody responses. Low doses of ovalbumin induced a state of tolerance that could be reversed by cyclophosphamide and primarily affected cell-mediated responses. Cyclophosphamide is believed to abrogate active suppression. It appears that these studies were delineating components of active suppression vs. anergy depending on dose. In addition, Hanson and Miller [61] reported two components of oral tolerance following oral administration of ovalbumin. They found tolerance was observed both in cyclophosphamide-treated and untreated animals but they were unable to transfer tolerance from cyclophosphamide-treated animals.

Our studies of oral tolerance in autoimmune models have found active suppression to be a primary mechanism and have identified regulatory cells generated following oral tolerance which act via the secretion to antigen-non-specific downregulatory cytokines following triggering by the fed antigen [62, 110]. These cells have been characterized in the rodent models of EAE orally tolerized to myelin basic protein (MBP). The regulatory cells can be either CD4+ or CD8+ [63] and act via the secretion of TGF-β following antigen-specific triggering [51, 110]. They transfer suppression in vivo and can suppress in vitro. While the epitopes of MBP triggering CD8+ regulatory cells following orally administered MBP are different than the encephalitogenic determinant [64] CD4+ TGF-β secreting regulatory cells may have structurally identical TCRs to encephalithogenic clones. TGF-β-secreting regulatory cells can also be found in Peyer's patches 24–48 h after one feeding of MBP [65]. Of note is that cells from Peyer's patches removed after one feeding of MBP do not proliferate in response to MBP even though they release TGF-β upon in vitro stimulation. The mechanism by which these regulatory cells are induced remains unknown. It is also not known the degree to which the generation of regulatory cells is related to unique antigen-presenting cells in the gut, the cytokine milieu, or other factors. In addition, ongoing studies from our laboratory in murine models of oral tolerance suggest that in addition to TGF-β secreting regulatory cells, the CD4+ regulatory cells secrete IL-4 and IL-10 in addition to TGF-β [110]. Understanding the biology of the regulatory cells generated by oral tolerance will be aided by cloning, which will require understanding the growth characteristics of such cells.

Antigen-Driven Bystander Suppression

Bystander suppression was first demonstrated in vitro when it was shown that cells from MBP-fed animals could suppress proliferation of an ovalbumin line across a transwell [66]. The cells from MBP-fed animals suppressed across the transwell only when triggered by the fed antigen. In an analogous fashion, cells from OVA-fed animals suppressed an MBP line across the transwell when stimulated with OVA. The soluble factor shown to be responsible for the suppression was TGF-β. Bystander suppression was then demonstrated in vivo. Feeding ovalbumin has no effect on MBP-induced EAE in the Lewis rat. However, if animals are fed ovalbumin and then given aqueous ovalbumin in the footpad following immunization in the footpad with MBP/CFA, EAE is suppressed. Suppression is mediated by OVA-specific regulatory cells which migrate to the draining lymph node and secrete TGF-β upon encountering OVA and thus inhibit the generation of the MBP-specific immune response being generated in the lymph node. Bystander suppression is specific to the fed antigen and its transferable. Further demonstration of bystander or tissue-specific suppression in vivo was obtained using MBP peptides. In the Lewis rat, MBP peptide 21–40 is a nonencephalitogenic epitope whereas 71–90 is the encephalitogenic epitope [64]. Peptide 21–40 triggers TGF-β release following oral tolerization and orally administered 21–40 suppresses 71–90 induced EAE in the Lewis rat. Furthermore, in 71–90 immunized animals protected by oral administration of peptide 21–40, DTH responses in the ear to peptide 71–90 are not suppressed whereas DTH responses to whole MBP are suppressed and suppression occurs because the 21–40 epitope is present in whole MBP to trigger TGF-β-secreting cells. Another example of tissue-specific by-stander suppression is the suppression of PLP-peptide induced disease in the SJL mouse by feeding myelin basic protein or MBP peptides [67].

Anergy

Anergy has only recently been demonstrated as a possible mechanism for oral tolerance [48, 49]. Anergy is defined as a state of T-lymphocyte unresponsiveness characterized by absence of proliferation, IL-2 production and diminished expression of IL-2R [68]. Anergy may be experimentally differentiated from clonal deletion by demonstrating the presence of antigen specific TcR clonotypes, or by release from the anergic state which is accomplished by pre-

Table 1

Animal model	Antigen fed
EAE (MS)	MBP, PLP
Arthritis (CII, AA)	type II collagen
Uveitis	S-antigen, IRBP
Diabetes (NOD mouse)	insulin, GAD
Myasthenia gravis	AChR
Thyroiditis	thyroglobulin
Transplantation	alloantigen, MHC peptide

culture of cells in IL-2 [69]. Under these conditions we have shown that a single feeding of 20 mg OVA induced a state of anergy in OVA specific T lymphocytes: cells did not respond to OVA by proliferation, OVA stimulation did not induce IL-2 production or IL2R expression, and the nonresponsive state was reversed by preculture of tolerized cells in IL-2 [49]. One other study has indirectly demonstrated anergy as a mechanism for oral tolerance [48]. Whitacre et al. [48] reported diminished IL-2 and IFNγ production in rats orally fed MBP in the presence of the soybean protease inhibitor; however, anergy was not confirmed in this study by TcR analysis or by IL-2-driven release. As discussed previously, the induction of anergy depends upon antigen dosage and frequency of feeding [50].

Treatment of Organ-Specific Autoimmune Diseases in Animals

Thompson and Staines [70] and Nagler-Anderson et al. [71] initially described suppression of collagen-induced arthritis by feeding type II collagen. Our laboratory [72–74] and that of Whitacre's [48, 75, 76] have studied suppression of EAE by orally administered myelin antigens. Additionally, we have investigated oral tolerance to suppress autoimmune models of uveitis [91–93], diabetes in the NOD mouse [77, 78] and adjuvant arthritis [79, 80], as well as orally administered alloantigen or MHC peptide in transplantation models [81–86]. Investigators have also demonstrated suppression of other autoimmune models by orally administered antigen (for review, please see Weiner et al. [87] and table 1). Thus, the ability to suppress autoimmunity in animal models via oral tolerance has been established although the mechanisms responsible may differ depending on the laboratory and the models being studied.

EAE

Orally administered guinea pig MBP suppresses EAE in the Lewis rat model [72, 75]. In a series of experiments we have shown this to be mediated by antigen-specific CD8+ T cells that act via the secretion of TGF-β [51, 63, 64]. Whitacre's group has also shown suppression of EAE in the Lewis rat with guinea pig MBP but have found clonal anergy to be the primary mechanism [48, 75, 76]. As discussed previously, it appears that the dosage of antigen fed accounts for the differences. In our experiments, a total of 3–5 mg was fed in 1-mg doses whereas Witacre's group fed 20 mg given in 5-mg doses plus soybean trypsin inhibitor (STI) which prevented degradation of antigen in the gut. Of note is that Whitacre also used male rats which have a more severe form of EAE and thus may be more difficult to tolerize orally. Although mechanisms of tolerization may have been different, both groups find suppression of disease, decreased proliferative responses, and decreased inflammatory responses in the brain.

Chronic relapsing EAE in the Lewis rats and strain 13 guinea pig was also suppressed by oral administration of MBP or a bovine myelin preparation that is also being administered in human clinical trials. There was no exacerbation of disease in these animals demonstrating that orally administered antigens do not appear to prime rather than suppress in an already immunized animal. Of note in the guinea pig model is that 10 mg of bovine myelin fed 3 times per week over a 3-month period suppressed disease and histologic manifestations whereas feeding 50 mg of bovine myelin did not. This may indicate that bystander suppression was responsible for the effect on chronic disease and that in some instances the oral administration of too high a dose will not suppress autoimmune models. As discussed later, loss of protection by orally administered antigen at higher doses was also seen with orally administered collagen to suppress adjuvant arthritis and orally administered insulin to suppress diabetes in the NOD mouse.

Detailed immunohistology was performed in animals orally tolerized with MBP and in animals naturally recovering from EAE [81]. Brains from OVA fed animals at the peak of disease showed perivascular infiltration with activated mononuclear cells which secreted the inflammatory cytokines IL-1, IL-2, TNFα, IFNγ, IL-6 and IL-8. Inhibitory cytokines TGF-β and IL-4 and prostaglandin E_2 (PGE_2) were absent. In MBP orally tolerized animals, there was a marked reduction of the perivascular infiltrate and downregulation of all inflammatory cytokines. In addition, there was upregulation of the inhibitory cytokine TGF-β. When bacterial lipopolysaccharide (LPS) was fed in addition to MBP, protection against EAE was enhanced and was associated with elevated IL-4 and PGE_2 in the brain [81, 88]. In control recovering animals (day 18), staining for inflammatory cytokines was diminished and there was upregulation of TGF-β and IL-4. These results suggest that the suppression of EAE, by oral tolerization and natural recovery, is related to regulatory cells that secrete inhibitory cytokines at the target organ. Prior feeding of MBP protects against actively induced EAE in the Lewis rat although it does not protect from adoptively transferred EAE [48, 89]. Nonetheless, spleen cells from orally tolerized animals will suppress adoptively transferred EAE when cotransferred with encephalitogenic cells or when injected into recipient animals at a different site at the time encephalitogenic cells are transferred [89]. This suppression was mediated by CD8+ T cells. Presumably, prior oral tolerization does not prevent adoptively transferred disease as the adoptively transferred cells migrate rapidly to the brain and initiate the inflammatory response before sufficient numbers of orally induced regulatory cells appear in the brain to downregulate the response. However, if regulatory cells from orally tolerated animals are transferred into the animal at the same as the encephalitogenic cells they migrate to the brain simultaneously, are activated by MBP to release factors such as TGF-β to prevent encephalitogenic cells from initiating the inflammatory response.

Epitopes of MBP distinct from the encephalitogenic region mediate suppression following oral tolerization. We tested the ability of overlapping 20 amino-acid peptides from MBP to trigger suppression mediated by spleen cells from Lewis rats orally tolerized to MBP. Spleen cells from MBP orally tolerized animals stimulated by residues 21–40, 51–70 and 101–120 of MBP suppressed proliferative responses of an OVA-specific cell line in a transwell system. This suppression correlated with secretion of TGF-β by cells stimulated with the peptide. In addition, T cells from animals fed the tolerogenic peptide 21–40 alone secreted TGF-β whereas no TGF-β release or in vitro suppression was observed in animals fed the MBP encephalitogenic determinant 71–90. As discussed previously, oral administration of peptide 21–40 suppressed disease induced by peptide 71–90. Of note is that orally administered 71–90 also suppressed disease. Studies are underway to determine whether this represents clonal anergy or the generation of a non-TGF-β-secreting regulatory cell. In studies of a variety of species, it was found that suppression of EAE was most effective when MBP from the homologous species was administered orally [74]. This has important implications for treatment of human autoimmune diseases. Human recombinant MBP has been cloned and expressed and is effective in suppressing EAE in the SJL mouse model [90].

In order to better delineate mechanisms of oral tolerance in the EAE model, a series of experiments have been initiated in the SJL model of EAE. We have found suppression of MBP- and PLP-induced disease by orally administered MBP or PLP and have also found suppression by the administration of MBP peptides. Ongoing studies have demonstrated both CD4+ and

CD8+ regulatory cells following oral tolerance in the murine model and work is in progress in an attempt to clone and characterize these cells.

Collagen and Adjuvant Arthritis

Previous investigators have demonstrated suppression of collagen-induced arthritis by feeding collagen type II [70, 71]. We have studied adjuvant arthritis (AA) in the rat, a well-characterized and more fulminant form of experimental arthritis [79]. Oral administration of chicken collagen type II (CII), given at a dose of 3 μg per feeding consistently suppressed the development of AA. A decrease in DTH responses to CII was observed that correlated with suppression to AA. Oral administration of collagen type I also suppressed AA; only minimal effects were seen with collagen type III. Suppression was antigen specific in that feeding collagen type II did not suppress EAE, and feeding MBP did not suppress AA. Suppression of AA could be adoptively transferred by T cells from CII-fed animals and was observed when CII was fed after disease onset. Of note is that suppression was observed at doses of 3 and 30 μg, but not at 300 or 1,000 μg. These results suggest that oral collagen is suppressing AA via bystander suppression rather than clonal anergy since active suppression may be lost at higher doses. The effectiveness of such small amounts of oral collagen may be related to the fact that collagen has repeating amino acid subunits.

Uveitis

Oral administration of S-antigen (S-Ag), a retinal autoantigen that induces experimental allergic uveitis (EAU), prevented or markedly diminished the clinical appearance of S-Ag-induced disease as measured by ocular inflammation [91–97]. Furthermore, oral administration of S-Ag also markedly diminished uveitis induced by the uveitogenic M and N fragments of the S-Ag. Oral administration of S-Ag did not prevent MBP-induced EAE. In vitro studies demonstrated a significant decrease in proliferative responses to the S-Ag in lymph node cells draining the site of immunization from fed versus nonfed animals. Furthermore, the addition of splenocytes from S-Ag-fed animals to cultures of a CD4+ S-Ag-specific lymphocyte line profoundly suppressed the line's response to the S-Ag, whereas these splenocytes had no effect on a PPD-specific lymphocyte line. The antigen-specific in vitro suppression was blocked by anti-CD8 antibody, demonstrating that suppression was dependent on CD8+ T lymphocytes. As in EAE, EAU was also suppressed by feeding S-Ag-related peptides that were either uveitogenic, cross-reactive or synthetic [93–97]. Gregerson et al. [96] using high and low doses of S-Ag peptides also found that low doses of antigen favors suppression whereas high doses induce unresponsiveness or anergy.

Diabetes

NOD diabetic mice spontaneously develop an autoimmune form of diabetes associated with insulitis. To test oral tolerance in the NOD model, we administered porcine insulin at a dose of 1 mg orally twice a week for 5 weeks and then weekly until 1 year of age [98]. The severity of lymphocytic infiltration of pancreatic islets was reduced by oral administration of insulin and there was a delay in the onset of diabetes. A decreased incidence of diabetes was seen in animals followed for 1 year. Suppression of insulitis as observed at a dose of 1 but not 5 mg. As expected, orally administered insulin had no metabolic affect on blood glucose levels. Furthermore, splenic T cells from animals treated orally with insulin adoptively transferred protection against diabetes, demonstrating that oral insulin generates active cellular mechanisms that suppress disease. Ongoing studies have demonstrated the ability to suppress insulitis by administering insulin peptides, the B chain of insulin, or GAD and immunohistochemical studies have demonstrated an increase of IL-4 in the islets of insulin-fed animals. Given the mechanism of antigen-driven bystander suppression, our results do not implicate autoreactivity to insulin as a pathogenic mechanism in the NOD mouse. Indeed, initial experiments suggest that orally administered glucagon can suppress insulitis.

Treatment of Autoimmune Diseases in Humans

The first attempts of oral tolerization may have been utilized by Native Americans who were thought to have fed their children *Rhus* leaves to prevent them from becoming sensitized to poison ivy [99]. Investigators have shown that exposure of a contact-sensitizing agent via the mucosa prior to subsequent skin challenge led to unresponsiveness in a portion of the subjects studied [100]. In another study on human volunteers, serum antibodies to bovine serum albumin (BSA) were measured before and after feeding large amounts of this antigen (0.1–1.5 mg of BSA per pound per day). Those subjects that had anti-BSA antibodies prior to eating BSA showed a rise in their serum anti-BSA titers. A similar response was observed when some subjects were given an injection of BSA. Subjects who did not have anti-BSA antibodies before or after the test did not respond to subsequent intradermal immunization [101]. Oral desensitization has also been attempted in Rh disease [102]. In an attempt at oral immunization, human volunteers were given capsules containing killed *Streptococcus mutans* and circulating IgA-producing cells were found in some subjects [103]. This suggests that some generation of a secretory immune response can occur following oral ingestion of microbial antigens. Orally administered KLH, 50 mg given daily for 2 weeks over a 3-week period has been reported to decrease subsequent cell-mediated immune responses although antibody responses were not affected [104]. In addition, preliminary experiments in subjects fed KLH have suggested that cell lines may be generated that suppress proliferative response although more work is needed in this area [105].

Multiple Sclerosis

In order to determine whether orally ingested autoantigens could affect the clinical course and immune responses in patients with an autoimmune disease, 15 patients with relapsing remitting multiple sclerosis were fed a capsule containing 300 mg of bovine myelin or placebo daily for 1 year [106]. Results demonstrated a decrease in MBP-reactive cells in the bloodstream of MS patients as compared to controls. There was no evidence of sensitization either as measured by antibody levels to MBP or PLP or by increased proliferative responses to the fed antigens at a 1-year period. Clinical responses demonstrated that 12 of 15 placebo-fed patients had major MS attacks whereas only 6 of 15 in the control group had attacks (p = 0.06). It appeared that a subgroup of patients that were either males or DR2$^-$ preferentially responded to the oral tolerization. However, the sample size was small and the degree to which this subgroup response will occur in future studies is unknown. Based on these observations, a multicenter <500 patient double-blind placebo-controlled trial of bovine myelin in DR2$^-$ relapsing remitting MS patients, both males and females, has begun. Of note is that trials of myelin basic protein given by multiple subcutaneous injections result in prominent immune responses to the antigen [107].

Rheumatoid Arthritis

A 60-patient double-blind trial of oral collagen administration to patients with rheumatoid arthritis demonstrated a decrease in joint swelling and disease index in patients fed collagen compared to placebo controls [108]. Patients were those previously or currently on immunosuppressant drugs such as methotrexate that had failed such therapy. Patients were taken off these medications and treated for a 3-month period. In the first month they received 100 μg of oral collagen per day and in the second and third months 500 μg per day. These doses were extrapolated from the small amounts of collagen used to suppress adjuvant arthritis in the Lewis rat. There were no toxicities or evidence of sensitization to type II collagen in fed patients as measured by anticollagen antibodies. There was no linkage to either DR type or sex in the patients that responded. In patients treated with the collagen there was less need for narcotic use during the course of the study and 4 patients in the collagen-treated group apparently had complete remission of their rheumatoid arthritis. Given the biologic effects seen, future trials will focus on other disease categories, more prolonged administration, and dosing studies. A multicenter trial is currently underway. Given what is known of the mechanism of oral tolerization, these studies do not establish that type II collagen is a target autoantigen in the disease. Indeed, given the low doses fed, it is possible that the effect may have been mediated by regulatory cells that migrated to the joint and released anti-inflammatory cytokines such as TGF-β or IL-4. Whether patients that went into complete remission represent a separate category remains to be determined.

Uveitis

An open-label pilot study has been performed on 2 patients with uveitis, 1 with Bechet's disease [109] and the other with pars planitis. In this open-label trial, patients had required steroids and/or cyclosporin to maintain visual function. Patients were started on 30 mg of bovine S-antigen three times a week and then tapered from steroids and immunosuppressive medication. A positive therapeutic response was observed in both patients in that over a 2-year period they were able to reduce their previous medication without worsening of vision and with decrease in S-antigen responsiveness. A double-masked placebo-controled trial of 45 patients is currently in progress.

Future Directions

Given the results in animal models of autoimmunity and initial studies in human disease states, it appears that orally administered autoantigens may find a place for the treatment of human organ-specific inflammatory autoimmune diseases. Such therapy would have the advantages of being orally administered, nontoxic, and antigen-specific. The mechanism of bystander suppression solves a major problem related to designing antigen or T-cell specific therapy of inflammatory autoimmune diseases since one needs not necessarily identify the target autoantigen for oral tolerance to be effective. As discussed above, it is likely that in human autoimmune disease states there are reactivities to multiple autoantigens from the target organ and multiple epitopes given that humans are an outbred population. Dosing appears to be important for stimulating the active suppression component of oral tolerance and identification of regulatory cells in humans following oral tolerization is critical for demonstrating the immunologic effects of oral tolerization. Given the results in animal studies, one would predict that homologous protein and the use of synergists or enhancers would increase the biologic efficiency of oral tolerance. In this regard, recombinant human proteins and the concomitant administration of immune adjuvants to enhance generation of regulatory cells would be required.

References

1   Adams CW, Poston RN, Buk SJ: Pathology, histochemistry and immunocytochemistry of lesions in acute multiple sclerosis. J Neurol Sci 1989;92:291–306.

2   Prineas JW, Wright RG: Macrophages, lymphocytes, and plasma cells in the perivascular compartment in chronic multiple sclerosis. Lab Invest 1978;38:409–421.

3   Prineas JW, Raine CS: Electron microscopy and immunoperoxidase studies of early multiple sclerosis lesions. Neurology 1976;26:29–32.

4   Woodroofe MN, et al: Immunocytochemical characterisation of the immune reaction in the central nervous system in multiple sclerosis. Possible role for microglia in lesion growth. J Neurol Sci 1986;74:135–152.

5   Hofman FM, et al: Immunoregulatory molecules and IL-2 receptors identified in multiple sclerosis brain. J Immunol 1986;136:3239–3245.

6   Hofman FM, et al: Tumor necrosis factor identified in multiple sclerosis brain. J Exp Med 1989;170:607–612.

7   Harding CV, Unanue ER: Quantitation of antigen-presenting cell MHC class II/peptide complexes necessary for T-cell stimulation. Nature 1990;346:574–576.

8   Adorini L, et al: In vivo competition between self peptides and foreign antigens in T-cell activation. Nature 1988;334:623–625.

9   Hickey WF, Kimura HP: Perivascular microglial cells of the CNS are bone marrow-derived and present antigen in vivo. Science 1988;239:290–292.

10 Fontana AW, Fierz W, Wekerle H: Astrocytes present myelin basic protein to encephalitogenic T-cell lines. Nature 1984;307:273–276.

11 Fierz W, et al: Astrocytes as antigen-presenting cells. I. Induction of Ia antigen expression on astrocytes by T cells via immune interferon and its effect on antigen presentation. J Immunol 1985;134:3785–3793.

12 Lee SC, Raine CS: Multiple sclerosis: Oligodendrocytes in active lesions do not express class II major histocompatibility complex molecules. J Neuroimmunol 1989;25:261–266.

13 McCarron RM, et al: Presentation of myelin basic protein by murine cerebral vascular endothelial cells. J Immunol 1985;143:3100–3103.

14 Panitch HS, Hafler DA, Johnson KP: Antibodies to myelin basic protein in cerebrospinal fluid in patients with multiple sclerosis; in Bauer HJ, Poser S, Ritter G (eds): Progress in Multiple Sclerosis Research. Heidelberg, Springer, 1980, pp 98–105.

15 Warren KG, Catz I: A myelin basic protein antibody cascade in purified IgG from cerebrospinal fluid of multiple sclerosis patients. J Neurol Sci 1990;96:19–27.

16 Wajgt A, Gorny M: CSF antibodies to myelin basic protein and to myelin-associated glycoprotein in multiple sclerosis. Evidence of the intrathecal production of antibodies. Acta Neurol-Scand 1983;68:337–343.

17 Lisak RP, Zweiman B: In vitro cell-mediated immunity of cerebrospinal fluid lymphocytes to myelin basic protein in primary demyelinating diseases. N Engl J Med 1977;297:850–853.

18 Johnson D, et al: Cell-mediated immunity to myelin-associated glycoprotein, proteolipid protein, and myelin basic protein in multiple sclerosis. J Neuroimmunol 1986;13:99–108.

19 Schluesener HJ, Wekerle H: Autoaggressive T lymphocyte lines recognizing the encephalitogenic region of myelin basic protein: In vitro selection from unprimed rat T lymphocyte populations. J Immunol 1985;135:3128–3133.

20 Martenson RE, Levine S, Sowniski R: The location of regions in guinea pig and bovine myelin basic proteins which induce experimental allergic encephalomyelitis in Lewis rats. J Immunol 1975;114:592–595.

21 McFarlin DE, Blank SE, Kibler RF: Experimental allergic encephalomyelitis in the rat: Response to encephalitogenic proteins and peptides. Science 1973;179:478–480.

22 Vandenbark AA, Hashim GA, Celnik B: Determinants of human myelin basic protein that induce encephalitogenic T cells in Lewis rats. J Immunol 1989;143:3512–3516.

23 Ota K, et al: T-cell recognition of an immunodominant myelin basic protein epitope in multiple sclerosis. Nature 1990;346:183–187.

24   Martin R, Howell MD, Jaraquemada D: A myelin basic protein peptide is recognized by cyto-
      toxic T cells in the context of four HLA-DR types associated with multiple sclerosis. J Exp Med
      1991;173:19–24.
25   Jaraquemada D, et al: HLA-DR2a is the dominant restriction molecule for the cytotoxic T cell
      response to myelin basic protein in DR2Dw2 individuals. J Immunol 1990;145:2880–2885.
26   Zhang JW, et al: Preferential peptide specificity and HLA restriction of myelin basic protein-
      specific T cell clones derived from MS patients. Cell Immunol 1990;129:189–198.
27   Wucherpfennig KW, Weiner HL, Hafler DA: T cell recognition of myelin basic protein. Immu-
      nol Today 1991;12:277–282.
28   Chou YK, et al: Response of human T lymphocyte lines to myelin basic protein: Association
      of dominant epitopes with HLA class II restriction molecules. J Neurosci Res 1989;23:
      207–216.
29   Pette M, et al: Myelin autoreactivity in multiple sclerosis: Recognition of myelin basic protein
      in the context of HLA-DR2 products by T lymphocytes of multiple-sclerosis patients and
      healthy donors. Proc Natl Acad Sci USA 1990;87:7968–7972.
30   Olsson T: Antimyelin basic protein and antimyelin antibody-producing cells in multiple
      sclerosis. Ann Neurol 1990;27:132–136.
31   Olsson T, et al: Autoreactive T lymphocytes in multiple sclerosis determined by antigen-in-
      duced secretion of interferon-gamma. J Clin Invest 1990;86:981–985.
32   Link H, et al: Persistent anti-myelin basic protein IgG antibody response in multiple sclerosis
      cerebrospinal fluid. J Neuroimmunol 1990;28:237–248.
33   Wucherpfennig KW, et al: Shared human T cell receptor V beta usage to immunodominant re-
      gions of myelin basic protein. Science 1990;248:1016–1019.
34   Ben Nun A, Liblau RS, Cohen L: Restricted T-cell receptor V beta gene usage by myelin basic
      protein-specific T-cell clones in multiple sclerosis: Predominant genes vary in individuals. Proc
      Natl Acad Sci USA 1991;88:2466–2470.
35   Kotzin BL, Karuturi S, Chou YK: Preferential T cell receptor B-chain variable gene use in
      myelin basic protein-reactive T cell clones from patients with multiple sclerosis. Proc Natl Acad
      Sci USA 1991;88:9161–9165.
36   Martin R, et al: A myelin basic protein peptide is recognized by cytotoxic T cells in the context
      of four HLA-DR types associated with multiple sclerosis. J Exp Med 1991;173:19–24.
37   Wucherpfennig KW, et al: Clonal expansion and persistence of human T cells specific for an im-
      munodominant myelin basic protein peptide. J Immunol 1994;in press.
38   Allegretta M, et al: T cells responsive to myelin basic protein in patients with multiple sclerosis.
      Science 1990;247:718–721.
39   Fujinami RS, Oldstone MBA: Amino acid homology between the encephalitogenic site of mye-
      lin basic protein and virus: Mechanism for autoimmunity. Science 1985;230:1043–1046.
40   Pullen AM, et al: Identification of the region of T cell receptor β chain that interacts with the
      self-superantigen Mls-1^a. Cell 1990;61:1365–1374.
41   Weiss A, et al: Role of T3 surface molecules in human T cell activation: T3-dependent activa-
      tion results in an increase in cytoplasmic free calcium. Proc Natl Acad Sci USA 1984;81:4169.
42   Meuer SC, Hussey RE, Fabbi M: An alternative pathway of T-cell activation: A functional role
      for the 50 kd T11 sheep erythrocyte receptor protein. Cell 1984;36:897–906.
43   Hunig TG, et al: Alternative pathway activation of T cells by binding of CD2 to its cell-surface
      ligand. Nature 1987;326:298–301.
44   Brod SA, et al: T-T cell interactions are mediated by adhesion molecules. Eur J Immunol 1990;
      20:2259–2268.
45   Kroemer G, Martinez AC: Mechanisms of self tolerance. Immunol Today 1992;13:401–404.
46   Miller JFAP, Moraham G: Peripheral T cell tolerance. Ann Rev Immunol 1992;10:51–70.
47   Mowat AM: The regulation of immune responses to dietary protein antigens. Immunol Today
      1987;8:93–98.
48   Whitacre CC, et al: Oral tolerance in experimental autoimmune encephalomyelitis. III. Evi-
      dence for clonal anergy. J Immunol 1991;147:2155–2163.
49   Melamed D, Friedman A: Direct evidence for anergy in T lymphocytes tolerized by oral admin-
      istration of ovalbumin. Eur J Immunol 1993;23:935–942.

50   Friedman A, Weiner HL: Induction of anergy and/or active suppression in oral tolerance is determined by frequency of feeding and antigen dosage. FASEB J 1993;in press.

51   Miller A, et al: Suppressor T cells generated by oral tolerization to myelin basic protein suppress both in vitro and in vivo immune responses by the release of TGFβ following antigen specific triggering. Proc Natl Acad Sci USA 1992;89:421–425.

52   Mowat AM, et al: Immunological response to fed protein antigens in mice. I. Reversal of oral tolerance to ovalbumin by cyclophosphamide. Immunology, 1982;45:105–113.

53   MacDonald TT: Immunosuppression caused by antigen feeding. I. Evidence for the activation of a feedback suppressor pathway in the spleens of antigen-fed mice. Eur J Immunol 1982;12: 767–773.

54   Guatam SC, Chikkala NF, Battisto JR: Oral administration of the contact sensitizer trinitrochlorobenzene: Initial sensitization and subsequent appearance of a suppressor population. Cell Immunol 1990;125:437–438.

55   Cowdery JS, Johlin BJ: Regulation of the primary in vitro response to TNP-polymerized ovalbumin by T suppressor cells induced by ovalbumin feeding. J Immunol 1984;132:2783–2789.

56   Richman LK, et al: Enterically induced immunological tolerance. I. Induction of suppressor T lymphocytes by intragastric administration of soluble proteins. J Immunol 1978;121: 2429–2433.

57   Strobel S, et al: Immunological responses to fed protein antigens in mice. II. Oral tolerance for CMI is due to activation of cyclophosphamide-sensitive cells by gut-processed antigen. Immunology 1983;49:451–456.

58   Miller S, Hanson D: Inhibition of specific immune responses by feeding protein antigens. IV. Evidence for tolerance and specific active suppression of cell-mediated immune responses to ovalbumin. J Immunol 1979;123:2344.

59   Bloom BR, Modlin RL, Salgame: Stigma variations: Observations on suppressor T cells and leprosy. Annu Rev Immunol 1992;10:453–488.

60   Sercarz E, Krzych U: The distinctive specificity of antigen-specific suppressor T cells. Immunol Today 1991;12:111–118.

61   Hanson DG, Miller SD: Inhibition of specific immune responses by feeding protein antigens. V. Induction of the tolerant state in the absence of specific suppressor T cells. J Immunol 1982; 128:2378–2381.

62   Weiner HL, et al: Suppression of organ-specific autoimmune diseases by oral administration of autoantigens. Progress in Immunol, vol VIII: 8th Int Congr Immunol, Budapest, 1992, pp 627–634.

63   Lider O, et al: Suppression of experimented autoimmune encephalomyelitis by oral administration of myelin basic protein. II. Suppression of disease and in vitro immune responses is mediated by antigen-specific CD8+ T lymphocytes. J Immunol 1989;174:791–798.

64   Miller A, Prabhu-Das M, Weiner, HL: Epitopes of myelin basic protein (MBP) that trigger TGF-β release following oral tolerization to MBP are different from encephalitogenic epitopes. FASEB J 1992;6:1686.

65   Santos LMB, et al: Oral tolerance to myelin basic protein induces TGF-β secreting T cells in Peyer's patches. J Immunol 1993;150:115A.

66   Miller A, Lider O, Weiner HL: Antigen-driven bystander suppression following oral administration of antigens. J Exp Med 1991;174:791–798.

67   Al-Sabbagh A, et al: Orally administered myelin basic protein suppresses proteolipid-induced experimental autoimmune encephalomyelitis in SJL mouse. Eur J Immunol, in press.

68   Schwartz RH: A cell culture model for T lymphocyte clonal anergy. Science 1990;248:1349–1356.

69   DeSilva DR, Urdahl KB, Jenkins MK: Clonal anergy is induced in vitro by T cell receptor occupancy in the absence of proliferation. J Immunol 1991;147:3261–3267.

70   Thompson HSG, Staines NA: Gastric administration of type II collagen delays the onset and severity of collagen-induced arthritis in rats. Clin Immunol 1986;64:581–586.

71   Nagler-Anderson C, et al: Suppression of type II collagen-induced arthritis by intragastric administration of soluble type II collagen. Proc Natl Acad Sci USA 1986;83:7443–7446.

72   Higgins PJ, Weiner HL: Suppression of experimental autoimmune encephalomyelitis by oral administration of myelin basic protein and its fragments. J Immunol 1988;140:440–445.

73   Brod SA, et al: Suppression of experimental autoimmune encephalomyelitis by oral adminis-
      tration of myelin antigens. IV. Suppression of chronic relapsing disease in the Lewis rat and
      strain 13 guinea pig. Ann Neurol 1992;in press.
74   Miller A, et al: Suppression of experimental autoimmune encephalomyelitis by oral administra-
      tion of myelin basic protein. V. Hierarchy of suppression by myelin basic protein from different
      species. J Neuroimmunol 1992;39:243–250.
75   Bitar D, Whitacre CC: Suppression of experimental autoimmune encephalomyelitis by the oral
      administration of myelin basic protein. Cell Immunol 1988;112:364–370.
76   Fuller KA, Pearl D, Whitacre CC: Oral tolerance in experimental autoimmune encephalomye-
      litis: Serum and salivary antibody responses. J Neuroimmunol 1990;28:15–26.
77   Zhang JA, et al: Suppression of diabetes in NOD mice by oral administration of porcine insu-
      lin. Proc Natl Acad Sci USA 1991;88:10252–10256.
78   Bisaccia G, et al: Heterogeneity of human T lymphocytes to bind sheep red blood cells in multi-
      ple sclerosis patients and controls. Boll Ist Sieroter Milan 1978;56:603–608.
79   Zhang JZ, et al: Suppression of adjuvant arthritis in Lewis rats by oral administration of type II
      collagen. J Immunol 1990;145:2489–2493.
80   Birnbaum G, et al: A comparison of regulatory cells in spinal fluid and blood in patients with
      multiple sclerosis and other neurologic diseases. Neurology 1990;40:1785–1790.
81   Khoury SJ, Hancock WW, Weiner HL: Oral tolerance to myelin basic protein and natural re-
      covery from experimental autoimmune encephalomyelitis are associated with down-regulation
      of inflammatory cytokines and differential upregulation of TGF-β, IL-4 and PGE expression in
      the brain. J Exp Med 1992;46:1355–1364.
82   Sayegh MH, et al: Down-regulation of the immune response to histocompatibility antigen and
      prevention of sensitization by skin allografts by orally administered alloantigen. Transplanta-
      tion 1992;53:163–166.
83   Sayegh MH, et al: Induction of immunity and oral tolerance with polymorphic class II
      major histocompatibility complex allopeptides in the rat. Proc Natl Acad Sci USA 1992;89:
      7762–7766.
84   Bonitati J, Kinkel WR, Manning E: Cerebrospinal fluid immunoglobulins: Test modifications
      and results in multiple sclerosis letter. Clin Chem 1976;22:1234–1235.
85   Baxevanis CN, et al: Decreased expression of HLA-DR antigens on monocytes in patients with
      multiple sclerosis. J Neuroimmunol 1989;22:177–183.
86   Baxevanis CN, et al: Monocyte defects causes decreased autoMLR in multiple sclerosis pa-
      tients. Adv Exp Med Biol 1988;237:839–842.
87   Weiner HL, et al: Oral tolerance: Immunologic mechanisms and treatment of animal and hu-
      man organ specific autoimmune diseases by oral administration of autoantigens. Ann Rev Im-
      munol 1994;in press.
88   Khoury SJ, et al: Suppression of experimental autoimmune encephalomyelitis by oral adminis-
      tration of myelin basic protein. III. Synergistic effect of lipopolysaccharide. Cell Immunol 1990;
      131:302–310.
89   Miller A, et al: Suppression of experimental autoimmune encephalomyelitis by oral administra-
      tion of myelin basic protein. VI. Suppression of adoptively transferred disease and differential
      effects of oral versus intravenous tolerization. J Neuroimmunol 1993;in press.
90   Oettinger HF, et al: Biological activity of recombinant human myelin basic protein. J Neuroim-
      munol 1993;44:157–162.
91   Nussenblatt RB, et al: Inhibition of S-antigen induced experimental autoimmune uveoretinitis
      by oral induction of tolerance with S-antigen. J Immunol 1990;144:1689–1695.
92   Thurau SR, et al: Immunological suppression of experimental autoimmune uveitis. Fortschr
      Ophthalmol 1991;88:404–407.
93   Thurau SR, et al: Induction of oral tolerance to S-antigen induced experimental autoimmune
      uveitis by a uveitogenic 20mer peptide. J Autoimmunol 1991;4:507–516.
94   Singh VK, et al: Suppression of experimental autoimmune uveitis in rats by the oral administra-
      tion of the uveitopathogenic S-antigen fragments ar a cross-reactive homologous peptide.
      Cell Immunol 1992;139:81–90.

95 Vrabec TR, et al: Inhibition of experimental autoimmune uveoretinitis by oral administration of s-antigen and synthetic peptides. Autoimmunity 1992;12:175–184.

96 Gregerson D, Obritsch W, Donoso L: Suppression and clonal anergy play roles in oral tolerance and EAU. Invest Ophthalmol Vis Sci 1993;34(suppl):902.

97 Suh EDW, et al: Splenectomy abrogates the induction of oral tolerance in experimental autoimmune uveoretinitis. Curr Eye Res, in press.

98 Weiner HL, et al: Antigen-driven peripheral immune tolerance. Suppression of organ-specific autoimmune diseases by oral administration of autoantigens. Ann NY Acad Sci 1991;636:227–232.

99 Dakin R: Remarks on a cutaneous affection produced by certain poisonous vegetables. Am J Med Sci 1829;4:98–100.

100 Lowney ED: Immunologic unresponsiveness to a contact sensitizer in man. J Invest Dermatol 1968;51:411–417.

101 Korenblatt PE, et al: Immune responses of human adults after oral and parenteral exposure to bovine serum albumin. J Allergy 1968;41:226–235.

102 Gold WRJ, et al: Oral densensitization in Rh disease. Am J Obstet Gyecol 1983;146:980-981.

103 Czerkinsky C, et al: IgA antibody-producing cells in peripheral blood after antigen ingestion: Evidence for a common mucosal immune system in humans. Proc Natl Acad Sci USA 1987;84: 2449–2453.

104 Husby S, et al: Oral tolerance in humans. T cell but not B cell tolerance to a soluble protein antigen. 7th Int Congr Mucosal Immunol, Prague 1992.

105 Polanski M, et al: Oral tolerization to KLH in humans: generation of antigen-specific lines that suppress proliferative responses. J Immunol 1993;150:114A.

106 Weiner HL, et al: Double-blind pilot trial of oral tolerization with myelin antigens in multiple sclerosis. Science 1993;259:1321–1324.

107 Salk RJS: A study of myelin basic protein as a therapeutic probe in patients with multiple sclerosis. Hallpike JF, Adams CWM, Toutellotte WW (eds): Multiple Sclerosis. University Press 1983, pp 621–630.

108 Trentham DE, et al: Effects of oral administration of type II collagen on rheumatoid arthritis. Science 1993;261:1727–1730.

109 Nussenblatt RB, et al: The treatment of the ocular complications of Behcet's disease with oral tolerization. 6th Int Conf Behcet's Disease, 1993. Amsterdam, Elsevier, 1994, vol 265, pp 1237–1240.

110 Chen Y, Icuchroo U, Hafler DA, Weiner, HL: Regulatory T cell clones induced by oral tolerance: suppression of autoimmune encephalomyelitis. Science 1994; in press.

David A. Hafler, MD, Center for Neurologic Diseases, LMRC-113,
221 Longwood Avenue, Brigham and Women's Hospital, Boston, MA 02115 (USA)

Adorini L (ed): Selective Immunosuppression: Basic Concepts and Clinica Applications.
Chem Immunol. Basel, Karger, 1995, vol 60, pp 150–160

....

Treatment of Autoimmune Disease:
To Activate or to Deactivate?

Irun R. Cohen[1]

Department of Cell Biology, The Weizmann Institute of Science, Rehovot, Israel

An autoimmune disease is the mark of an immune system gone wrong. The autoimmune disease challenges clinical immunology to set the system right. In recent years, immunology has deciphered much of the basic chemistry of the cells, antibodies, antigen receptors, cytokines and cytokine receptors that comprise the immune system. This new molecular information is important because it should speed the design of therapies rationally targeted to critical elements in the pathologic autoimmune process. But to choose the most effective therapy, to foresee just what the novel therapies are likely to accomplish, we need to have some idea of the pathophysiology of autoimmunity: What is it that has gone wrong, what are we trying to fix? These questions are not mere philosophical niceties; they can guide concrete clinical decisions. And the strategies beyond the decisions may be poles apart. For example, some of the new therapies have been devised to inactivate immune responses by blocking MHC-T-cell receptor interactions or by killing specific T cells with toxin conjugates [1]. In contrast, other proposed therapies such as oral administration of antigen [2] or vaccinations with T cells or with peptides [3], have been devised to *activate* immune responses. The seemingly opposite strategies of *inactivation* and *activation* are each rational to their adherents because they rest on two different interpretations of autoimmunity: the clonal selection paradigm and the cognitive paradigm. The aim of this review is to articulate these paradigms and their implications for the therapy of autoimmune diseases.

---

[1] Incumbent of the Mauerberger Chair of Immunology and the Director of the Robert Koch-Minerva Center for Research in Autoimmune Diseases.

Clonal Selection

The clonal selection paradigm asserts that the echelons of lymphocytes that form the immune system are selected entirely by encounter with antigen [4]. This concept has wide currency and need not be described here. I will cite only two tenets of clonal selection relating to autoimmunity and autoimmune disease:

(1) The healthy immune system must not include lymphocytes that recognize self antigens; autoimmune lymphocytes, at least the T cells, are removed from the repertoire by processes of negative selection [5] and anergy [6] that take place in the thymus and in the periphery.

(2) Autoimmune diseases, therefore, can have no intrinsic regularity in their clinical or immunological expression; autoimmune diseases are caused by autoimmune lymphocytes that arise from random mutations or unstructured escape from deletion. At the risk of over-simplification, we can say that the clonal selection paradigm asserts that autoimmune disease develops because an entity that should be absent, an autoimmune lymphocyte clone, is present by accident. The failure of the immune system to have deleted the forbidden clone should then be rectified by blocking with various antibodies, for example, the clone's ability to recognize its antigen, or by medicinally killing the clone, for example by toxin conjugates targeted to activated T cells [1].

In other words, an autoimmune disease is caused, according to the clonal selection paradigm, by aberrant *activation* of the immune response [7]. The rational answer to harmful activation is to find a way to *deactivate* the pathogenic lymphocytes. Thus, the clonal selection paradigm argues for suppression of the immune system. Unlike the cytotoxic drugs and steroids now in use with their broad and uncontrollable effects, the new molecular immunology makes it feasible to design highly specific silver bullets that can target the forbidden clones of autoimmune lymphocytes and so de-activate the autoimmune agents of disease without inactivating the entire immune system.

The Cognitive Paradigm

Elsewhere, I have presented an alternative to the clonal selection paradigm called the cognitive paradigm [8, 9]. The cognitive paradigm takes account of two empirical observations that contradict the tenets of orthodox clonal selection relating to autoimmunity: (1) Healthy immune systems do in fact contain T cells and B cells that recognize certain dominant self antigens; some autoimmunity is natural. (2) Autoimmunity is not chaotic, but manifests quite predictable immunologic characteristics.

Healthy humans and healthy rodents bear natural autoantibodies and autoimmune T cells to similar sets of self antigens. Indeed, many of the same self antigens attacked in diseases are also recognized in health. Moreover, a small number of diseases encompasses the majority of autoimmunity patients: rheumatoid arthritis, type I diabetes mellitus, multiple sclerosis, systemic lupus erythematosus. Suffering a disease may be accidental; but the disease visited upon the patient is highly structured. This is true even across species: systemic lupus erythematosus and type I diabetes are quite similar in their immunology in both mice and humans. Order and predictability imply that natural autoimmunity and autoimmune diseases are governed by rules. But what function is served by having autoimmunity built into the system?

The Internal Physician and the Problem of Ambiguity

One way to think about the functions of the immune system is to consider it, metaphorically, as the body's internal physician, a physician specializing in infectious diseases (and perhaps in some types of cancer). This internal physician, like any external physician, performs two functions: diagnosis and treatment. The immune system has to diagnose the nature of a threat to the integrity of the body (virus, bacterium, higher parasite) and it has to prescribe the mix of therapeutic responses (types of T cells, B cells, antibodies, cytokines, inflammation) best suited to destroy the invader.

The real problem of the immune system, the one that makes simple clonal selection unworkable, is antigenic ambiguity: many of the key protein antigens of invading microbes and other parasites contain amino acid sequences that are identical to those of the host body. Molecular biology has demonstrated that evolution conserves useful genes. Nature does not reinvent the wheel each time she designs a new creature. Once evolution comes upon a molecule that performs an essential function, it conserves the molecule's genes for further use in other instances. Indeed the more important the molecule, the more its sequence is conserved across the epochs. Heat shock protein (hsp) molecules, essential for all cells, are a telling example: humans and bacteria manifest 50–80% identity of amino acid sequences in the various members of the hsp family, nevertheless, hsp molecules are recognized as dominant antigens of many infectious agents [10]. These conserved molecules are also among the self antigens recognized in healthy, natural autoimmunity [8, 9]. How does the immune system know how to respond appropriately to hsp and other molecules that antigenically are both self and not-self and that may accompany either health or infection? Recognition alone is not sufficient. The immune system must interpret the meaning of an antigen; to interpret, it has to resort to cognition.

Diagnostic Interpretation

The cognitive paradigm proposes that the immune system, the internal physician, interprets the meaning of antigens using a strategy similar to that used by an external physician to interpret the meaning of clinical signs. Interpretation is brought about using the cognitive process called *differential diagnosis*. Differential diagnosis is a way of making clinical decisions based on searching for a string of relevant information (history, physical examination, laboratory tests) and on organizing this information in the light of the physician's knowledge. The external physician will interpret the meaning of a fever, for example, according to what he knows about fever and what he learns about the patients. Diagnosis, in essence, is paradigmatic; it is the process of fitting the actual case to one of the archetypal models that the physician has in mind [11]. Implicit in the process are *memory*, the storage of information, and *learning*, the adjustment, through experience, of the archetypal models and the program of assembling information. The cognitive paradigm teaches that the immune system too interprets antigens by building internal representations (archetypal categories) and assembling strings of relevant signals (fig. 1).

Images of Self

Physicians know that internal images exist; they use mental images when they practice differential diagnosis. Cognitive scientists talk about internal representations of the external world that function in mental cognition [12], but nobody knows how these representations are actually encoded in the mind. Immunologists appear to be better informed; they know the molecular basis of at least some of the images in the immune system. The idea of immune internal images was made explicit by Jerne [13] in his anti-idiotypic theory of immune regulation. The antigen receptors of T and B cells are structurally complementary to the epitopes (sites) they recognize on their antigens. Thus, antigen receptors structurally are negative images of parts of antigens, just as locks and keys, in part, are negative images of each other. Anti-idiotypic antibodies that bind to the antigen-binding sites of other antibodies are structurally complementary to the negative images of antigens; thus, anti-idiotypic antibodies can be positive images of antigens (a negative of a negative can create a positive). The images created by T-cell idiotypes and anti-idiotypes are complicated by the fact that T-cell receptors see processed antigen fragments in the clefts of major histocompatibility complex (MHC) molecules; nevertheless, T-cell receptors are images all the same.

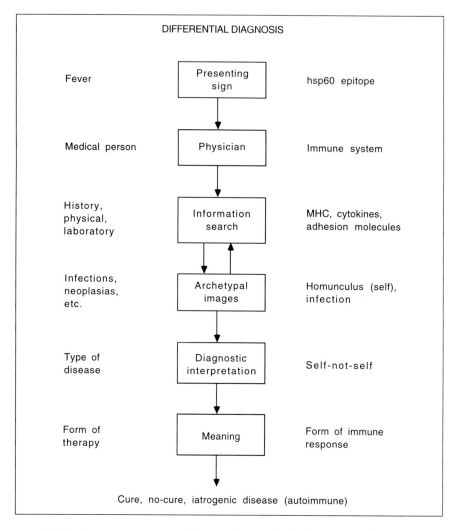

Fig. 1. The process of differential diagnosis – medical and immunological. The meaning of a presenting sign, such as a 'fever' for a medical person or an 'hsp60 epitope' for the
immune system, is decided in both cases by a search for the appropriate additional information needed to define the context of the sign. The person and the immune system are
each guided by internal archetypal images. The medical person's internal images of diseases are the accumulated useful experience of the profession taught in medical school
and modified by the person's actual experience. The immune system's internal images of
the self (the homunculus) and of infection in general are formed by the accumulated, evo-
lutionary experience of the species carried in the germ line combined with the individual's
somatic experience recorded in the antigen-receptor repertoire. The diagnostic interpretation of the person and of the system is the outcome of the interaction between the information search and the preformed archetypal images. The medical person decides whether

Because we humans are so attuned to visual information, we tend to think of images as organized pictures. Nevertheless, receptors on wandering cells or cell networks can also create images, albeit images that are distributed about the body.

Now, if the healthy immune system naturally contains antibodies, B cells and T cells capable of recognizing a standard set of self antigens, and if the system also contains anti-idiotypes to these elements, then we can state that the immune system contains both positive and negative internal images of particular self molecules. These images constitute what I call the *immunological homunculus* [8–10]. I borrowed the term homunculus from neurology where it refers to representations of the body encoded in the brain. Since self molecules are the first antigens the immune system meets, then it follows that the formation of the immunological homunculus precedes actual contact with foreign antigens. The T-cell imunological homunculus probably develops through the positive selection of some autoimmune T cells in the thymus, but much remains to be discovered about its origins; are particular self antigens programmed to be expressed in the thymus [14] to positively select autoimmune T cells? The homunculus formed by the natural autoantibodies is now being studied using a new technology [15].

It appears that many of the self molecules in the homunculus set, such as the hsp molecules, have been highly conserved in evolution and are shared by hosts and parasites [10]. It can be argued that autoimmunity to highly conserved molecules expressed both by host and by parasites might serve to prime the host immune response against the parasites. Note that tight regulation by anti-idiotypic and other regulatory cells is required to avoid the development of autoimmune diseases when the system transiently exploits natural autoimmunity to hsp and other conserved molecules to fight infection [10]. Thus, the immunological homunculus, regulated autoimmunity, is one of the ways evolved by the immune system to deal with the problem posed by conserved

the case fits any of the recognized disease categories in his stock of internal mental images. The immune system automatically interprets whether the sign relates to self or not-self by integrating the string of acquired information with the archetypal images. (Note that the distinction between the self and the not-self is a matter of interpretation. The self is defined by a set of interactions; it is not an immutable given.) The functional meaning of the presenting sign is the form of treatment prescribed by the medical person or the type of immune response deployed by the immune system. The diagnosis is correct if the therapeutic response cures the patient. A wrong diagnosis may result in no cure or, unfortunately, in an iatrogenic disease. In the case of the immune response, a wrong diagnosis may mean an autoimmune disease. Wrong diagnoses of both sorts can be corrected, according to the cognitive paradigm, by supplying the right information.

molecules shared by host and infectious agents, the problem of ambiguous self-not-self [16]. Of course, we must still explain the need for natural autoimmunity to self antigens that are not shared with parasites.

Images of Infections

In addition to the immunological homunculus, the immune system contains primordial internal images of infectious agents encoded by the many cell receptors specific for microbial products: components of bacterial cell walls, lipopolysaccharides, viral nucleic acids [8, 9]. Receptors for the cytokines triggered by infection can also be viewed as internal images of infection, images which reflect not the structure of infectious agents but their presence. These types of receptors for infectious agents are invariant; they are determined by germ-line genes and exemplify a primitive adaptation of the species to pervading features of infectious agents. Antibodies, B cells and T cells specific for microbial antigens are also internal images of the microbes, but these specific images are learned through the actual infectious experience of the individual. Thus, immune adaptation is expressed at two levels: a first, crude image of infection is encoded in the germ line; a second, refined image is encoded by the antigen-specific T and B cells activated in each individual through experience.

Immune Interpretation: Strings of Information

Table 1 lists four categories of information which the immune system uses to make diagnostic interpretations and therapeutic decisions. The first category (table 1) includes the receptor molecules that recognize antigens, what is usually called the repertoire of the immune system. These receptors are generated in each individual by random somatic processes. The original clonal selection paradigm focused its attention on this category of information, almost exclusively. The clonal selection paradigm was formulated early, at the time when immunologists still were unaware of the considerable resources of immune information encoded within the germ line: networks of cytokines, cell interactions, and antigen processing and presentation (table 1). Sections III and IV in table 1 include entities also unknown or disregarded by the formulators of clonal selection: the internal image of self (the immunological homunculus) and the primordial image of infection. The cognitive paradigm is a global view of the immune system up-dated to encompass the additional information that immunologists have discovered since the clonal selection paradigm.

Table 1. Immune information

I	Clonal recognition of antigen by way of somatic genetic recombination The repertoire of antibodies and antigen receptors on T and B cells
II	Germ-line recognition of context: accessory signals

I   Clonal recognition of antigen by way of somatic genetic recombination
      The repertoire of antibodies and antigen receptors on T and B cells

II   Germ-line recognition of context: accessory signals
    1   Cytokines [26]:
       Interleukins 1a, 1b, 2, 3, 4, 5, 6, 7, 8, 9, 10, 11, 12; TNFα, TNFβ, TGF-β1,
       TGF-β2, TGF-β3; interferons α, β, γ, etc.
    2   Ligand pairs in inter-leukocyte interactions [27]:
       MHC class I, class II; CD3, CD4, CD8, LAG-3, LFA-1, ICAM-1, CD2, LFA-3,
       CD28, B7, CTLA-4, CD40, CD40 ligand, CD45Ro, CD22, CD5, CD72; etc.
    3   Ligand pairs in leukocyte-tissue interactions [28]:
       Selectins (E, P, L) – carbohydrate ligands; IgCAM integrins; addressins; CD31;
       LFA-3-CD2; etc.
    4   Molecules involved in the processing and transport of antigens and
       antigen fragments

III   Primordial internal image of infection: receptors for microbial products [9]

IV   The immunological homunculus: natural autoimmunity [9]

Many of the elements listed in table 1 are membrane receptors connected directly or indirectly to protein kinases and other mechanisms of signal transduction. It is obvious from the large number of these receptors that lymphocytes are capable or receiving ancillary information from many sources in additon to recognizing antigens. It is the integration of this information which tells the lymphocytes what to do about the antigens it sees. Integration is interpretation; the nature of the response follows. For example, T cells of the Th1 type are involved in causing cell-mediated tissue damage; T cells of the Th2 type are involved in inducing antibody production by B cells. Whether a naive T cell becomes a Th1 or Th2 cell in response to a given antigen depends on many factors including the cytokine environment (IL-2 and interferon-γ vs. IL-4 and IL-10), the presence or absence of CD8 cytotoxic T cells, the nature of the antigen-presenting cell (macrophage vs. B cell), the concentration of the antigen and its mode of entry into the antigen-presenting cell (phagocytosis vs. pinocytosis) [17]. Important biological consequences depend on whether the immune system decides to generate Th1 or Th2 T cells to a particular antigen, but these are only two of many different possible effector agents that can result from antigen recognition.

Here lies the basic difference between the clonal selection and the cognitive paradigms: the former entrusts the antigens to organize the immune system; the latter proposes that the immune system has been programmed by evolution to interpret antigens, albeit automatically and unconsciously, like a very complex computer.

Autoimmune Disease: A Misinterpretation

The cognitive view of autoimmune diseases differs from that of the clonal selection paradigm. The autoimmune clones specific for the dominant self antigens are naturally present in the homunculus; nonetheless, these autoimmune clones do not normally cause disease for at least two reasons: the clones are tightly regulated by specific anti-idiotypic and other mechanisms, and the autoimmune clones see their target self antigens in contexts that do not incite an effector response (for example, in the absence of inflammatory cytokines, or in the presence of suppressive signals). The transition from natural, benign autoimmunity to autoimmune disease follows two circumstances: the self antigen is presented inadvertently in a context that drives a damaging effector response and the anti-idiotypic and other regulatory connections are too weak to counteract the stimulus and restore a healthy equilibrium [8–10].

The transition from benign, natural autoimmunity to an autoimmune disease has been studied in rodents in the disease called experimental autoimmune encephalomyelitis (EAE), which models aspects of multiple sclerosis (MS). EAE is caused by activated T cells with receptors for the self-antigen myelin basic protein (MBP) [18]. Anti-MBP T cells are also detectable in MS patients [19]. However, it has been discovered that potentially virulent anti-MBP T cells are present in healthy rats that will never develop EAE spontaneously [20], and also in healthy people who will probably never develop MS [21]. But clinical EAE can be induced readily by immunizing rats with MBP emulsified in oil containing dead mycobacteria, a material called complete Freund's adjuvant [18]. Apparently, the adjuvant supplies MBP with the ancillary signals indicative of an infection, and the naturally quiescent anti-MBP T cells are driven to differentiate into activated effector T cells, causing clinical EAE. Thus, the transition from benign to noxious autoimmunity in EAE and in other experimental diseases is induced by signals that dress MBP or the other self antigens in the context of infection [9].

An aberrant expression of MHC class II molecules has been proposed as a mechanism to explain the induction of clinical autoimmunity [7]. The cognitive paradigm would consider expression of MHC molecules as only one of many elements capable of defining the context of an antigen (table 1). Be that as it may, the transition to disease is the result of inappropriate contextual information.

Note, however, that the perpetuation of autoimmune disease differs from the induction of autoimmune disease. Despite the inciting context, the healthy immune system of the rat usually learns to resist the inciting context of MBP in adjuvant, probably by activating and amplifying the anti-idiotypic regulatory T cells inherent in the immunological homunculus [8–10]. EAE cannot be induced a second time, unless the regulatory T cells have been inhibited. Relapsing or progressive MS would also seem to require some insufficiency of the regulatory cells.

According to the cognitive paradigm, medicine might mimic nature by activating and strengthening the regulatory cells connected to the specific autoimmune lymphocytes through treatments such as T-cell vaccination, found to be effective in experimental animals [22] and recently in MS [23]. Another way to stimulate regulatory cells is by oral tolerance [2]. These simple forms of treatment would seem to work because, to avoid chaos, complex biological systems must organize themselves to focus on specific bits of information. Such bits of information function as regulatory elements because they influence the state of the system. The scope of this article does not allow a discussion of the self-organization of systems [24], but the power of small bits of information to control complex behavior is an observable fact of life: witness the response of the central nervous system to a certain smile, to a kind word. Noxious behaviour by the immune system too is susceptible to modification by communicating with the system using signals the system understands. Even a spontaneous autoimmune disease such as the diabetes of the NOD mouse can be treated by vaccination with a suitable T-cell clone or with an appropriate peptide [3, 25]. Thus, safer and more effective therapies for autoimmune diseases will emerge as immunology uncovers and exploits the critical bits of information that the immune system has been adapted to interpret cognitively. Like the nervous system, the immune system can learn to behave itself. Properly informed, the internal physician will make it right.

References

1   Bach JF: Immunosuppressive therapy of autoimmune diseases. Immunol Today 1993;14: 322–325.
2   Weiner HL, Mackin GA, Matsui M, Orav EJ, Khoury SJ, Dawson DM, Hafler DA: A double-blind pilot trial of oral tolerization with myelin antigens in multiple sclerosis. Science 1993;259: 1321–1324.
3   Elias D, Reshef T, Birk OS, van der Zee R, Walker MD, Cohen IR: Vaccination against autoimmune mouse diabetes with a T-cell epitope of the human 65-kDa heat shock protein. Proc Natl Acad Sci USA 1991;88:3088–3091.
4   Burnet FM: The Clonal Selection Theory of Acquired Immunity. Cambridge, Cambridge University Press, 1959.
5   Marrack P, Kappler JW: How the immune system recognizes the body. Sci Am 1993;269:49–55.

6   Schwartz RH: A cell culture model for T lymphocyte clonal anergy. Science 1990;248: 1349–1356.
7   Bottazzo JF, Pujol-Borrell R, Hanafusa T, Feldman M: Hypothesis: Role of aberrant HLA-DR expression and antigen presentation in the induction of endocrine autoimmunity. Lancet 1983;ii:1115–1119.
8   Cohen IR: The cognitive principle challenges clonal selection. Immunol Today 1992;13: 441–444.
9   Cohen IR: The cognitive paradigm and the immunological homunculus. Immunol Today 1992; 13:490–494.
10 Cohen IR, Young DB: Autoimmunity, microbial immunity, and the immunological homunculus. Immunol Today 1991;12:105–110.
11 Kassirer JP: Diagnostic reasoning. Ann Intern Med 1989;110:893–900.
12 Shanon B: Representations: senses and reasons. Philos Psychol 1991;4:355–374.
13 Jerne NK: Toward a network theory of the immune system. Ann Immunol (Paris) 1974;125C: 373–89.
14 Kourilsky P, Claverie JM, Prochnicka-Chalufour A, Spetz-Hagberg AL, Larsson-Sciard EL: How important is the direct recognition of the polymorphic MHC residues by TCR in the generation of the T-cell repertoire? Cold Spring Harbor Symp Quant Biol 1989;54:93–103.
15 Nobrega A, Haury M, Grandien A, Malanchère E, Sundblad A, Coutinho A: Global analysis of antibody repertoires. II. Evidence for specificity, self-selection and the immunological 'homunculus' of antibodies in normal serum. Eur J Immunol 1993;23:2851–2859.
16 Cohen IR: Kadishman's tree, Escher's angels and the immunological homunculus; in Coutinho A, Kazatchkine M (eds): Autoimmunity: Physiology and Disease. New York, Wiley-Liss, 1994, pp 7–18.
17 Fitch FWE, Mckisic MD, Lanck DW, Gajewski TF: Differential regulation of murine T lymphocyte subsets. Ann Rev Immunol 1993;11:29–48.
18 Ben-Nun A, Wekerle H, Cohen IR: The rapid isolation of clonable antigen-specific T lymphocyte lines capable of mediating autoimmune encephalomyelitis. Eur J Immunol 1981;11:195–199.
19 Ota K, Matsui M, Milford EL, Mackin GA, Weiner HL, Hafler DA: T-cell recognition of an immunodominant myelin basic protein epitope in multiple sclerosis. Nature 1990;346:183–187.
20 Mor F, Cohen IR: Shifts in the epitopes of myelin basic protein recognized by lewis rat T cells before, during and after the induction of experimental autoimmune encephalomyelitis. J Clin Invest 1993;92:2199–2206.
21 Burns J, Rosenzweig A, Zweiman B, Lisak RP: Isolation of myelin basic protein-reactive T-cell lines from normal human blood. Cell Immunol 1983;81:435–440.
22 Cohen IR: T-cell vaccination in immunological disease. J Intern Med 1991;230:471–477.
23 Zhang J, Medaer R, Stinissen P, Hafler D, Raus J: MHC-restricted depletion of human basic protein-reactive T cells by T cell vaccination. Science 1993;261:1451–1454.
24 Atlan H: Self-creation of meaning. Phys Scripta 1987;36:536–576.
25 Elias D, Cohen IR: Peptide therapy for diabetes in NOD mice. Lancet 1994;343:704–706.
26 Miyajima A, Kitamura T, Harada N, Yokata T, Arai K-I: Cytokine receptors and signal transduction. Annu Rev Immunol 1992;10:295–331.
27 Parker DC: T cell dependent B cell activation. Annu Rev Immunol 1993;11:331–360.
28 Bevilacqua MP: Endothelial-leukocyte adhesion molecules. Annu Rev Immunol 1993;11:767–804.

Prof. Irun R. Cohen, Department of Cell Biology, The Weizmann Institute of Science,
Rehovot 76100 (Israel)

Subject Index

blocking peptides 390
nicotinamide 33
objectives 32
rapamycin 36
T cell
  receptor
    antibodies 39
    Vβ inhibition 40
    vaccination 40
transfer by bone marrow transplant 32
Integrins
  antibodies, blocking of immune response
    24, 25
  role in lymphocyte adhesion 23, 24
Interleukin-1, receptor antagonist 52
Interleukin-2
  growth factor activity 93
  regulation by costimulatory signals 81
  reversal of tolerance 21, 22
Interleukin-4
  biological activity 54, 55
  role in rheumatoid arthritis 55, 56
Interleukin-10
  induction of diabetes mellitus 53
  role in rheumatoid arthritis 53

Major histocompatibility complex
  antibody, prevention of diabetes 38
  class I molecules
    antigen presentation 37
    blocking peptides 39
    peptide binding specificity 61
    transgenic mice 37, 38
  class II molecules
    blocking peptides
      CY-760.50 62, 63
      design 62, 63
      dose response 63, 64
      therapeutic potential 62–64, 94
    peptide binding
      affinity effect on cytokine secretion
        14, 15
      competition 11, 63, 127
      effect on Th cell differentiation
        14, 15
      specificity 61
  peptide-binding groove 80

structure 79, 80
T cell receptor recognition 80
Modified receptor ligands, see T cell
  receptor
Multiple sclerosis
  autoantigens
    reactivity 127, 128
    types 73, 127–129
  encephalitogenic T cells
    activation
      autoantigen 130, 131
      bystander CD2 pathway 132
      molecular mimicry 131, 132
      superantigens 132
    assay 128, 129
    frequency 103–105, 130, 131
    levels in controls 103, 130
    lymphokine secretion 112
    phenotype 106
    reactivity 128, 129
    regulation 120
    response to antigens 107, 108
    role in pathogenesis 100, 101, 112
    specificity 106
    V gene bias 108, 109, 121, 130
  etiology 126
  inflammation mechanism 126, 127
  initiation 102
  myelin basic protein processing 105,
    106, 127
  oral antigen therapy 143
  plaque formation 102, 126
  similarity to experimental allergic ence-
    phalomyelitis 102
  T cell receptor
    characterization 130
    peptide therapy
      characterization of reactive T cells
        117–119
      efficacy 117
Myelin basic protein
  cryptic determinants 4, 5, 7
  immunodominant regions 129
  induction of experimental allergic ence-
    phalomyelitis in animals 4, 8, 13
  inhibition of experimental allergic ence-
    phalomyelitis by analogs 71